Choose!
Life or Death

Reams Biological Theory of Ionization

By Carey A. Reams

NUTRITIONAL COUNSELORS OF AMERICA, INC.
HOLISTIC LABORATORIES, INC.
TAMPA, FLORIDA

First printing, April 1978 © New Leaf Press
Second printing, June 1978 © New Leaf Press
Third printing, February 1982 © Holistic Laboratories, Inc.
Fourth printing, July 1983 © Holistic Laboratories, Inc.
Fifth printing, December 1987 © Holistic Laboratories, Inc.
Sixth printing, 1990 © Holistic Laboratories, Inc.
Seventh printing, 1997 © Holistic Laboratories, Inc.

ISBN 0-9619345-0-6

Choose! Life or Death © 1982 by Holistic Laboratories, Inc. Copyright assigned to Holistic Laboratories in January 1982. All rights reserved. Printed in the United States of America. No part of this book may be used or reproduced in any manner whatsoever without written permission of the publisher except in the case of brief quotations in articles and reviews. For information write: NCA, Inc./Holistic Laboratories, Inc., June M. Wiles, President/CEO/Instructor — RBTI, P.O. Box 16793, Tampa, Florida 33687 or NCA, Inc., 1725 East Fowler Ave., Tampa, FL 33612 • (813) 977-1000/ (813) 988-3945.

CONTENTS

Preface		7
Publisher's Note		12
Acknowledgments		14
Chapter		
1	Equation Numbers	15
2	Discovery	24
3	Creation	35
4	Homo Sapiens	39
5	Some of the Chemistry of Foods	48
6	Application, Why and Wherefore	52
7	Cause of Diseases	62
8	Cause and Effect	70
9	Reams Biological Theory of Ionization	80
10	Range A-E	86
11	How a New Cell Is Developed	89
12	Testing Procedures	93
13	Low Blood Sugar	97
14	Obesity	103
15	Fads or Facts	109
16	Min-Col	117
17	Case Histories	121
18	RBTI and Pastoral Counseling	129
19	Oxygen	135
20	Digesting Food	141
21	The Greatest Love of My Life	149
22	The Man Carey A. Reams	155

Carey Reams

A TRIBUTE TO CAREY A. REAMS

Who is man that thou art mindful of him? God uses men for His purpose and according to his ability and willingness to serve Him. Man serves God in four ways: emotionally, physically, intellectually and spiritually. Carey A. Reams was willing to serve God in every way, but especially in the physical realm. Like Dr. Luke of the Bible, Dr. Reams chose to work with God and man in the absolutely natural (chemistry and mathematics) way, embracing the pure God-given laws of chemistry and physics. Though he lived daily with persecution, he was never discouraged and NEVER thought of giving up. HE HAD A JOB TO DO!

When I first learned of Carey Reams, I was operating a health food store (my own — Monroe's Health Foods) in Tampa, Florida. I had had five years of nutrition, was doing counseling and ever saying, "There is more, but what is it?" I said this to a vitamin representative out of Atlanta, who on her next visit brought me a book, *Health Guide for Survival* (now out of print) about Carey Reams. What I read caused me to burst out audibly, "This is it, this is it!" Since that day my whole life has changed because one man chose to use his talents to serve God — in the natural realm. I met Dr. Reams in 1975 when I went for two weeks to the retreat. I took my first class in December of 1975 and all successive classes after that (six, not nine). In 1977, while I was still in classes, I had Dr. Reams come to Florida to teach RBTI. By this time I had mastered the nutritional aspects of BTI counseling and combining our talents we gave every student "more than his money's worth." Dr. Reams and I taught classes for "my" people from 1977 to his untimely death on August 5, 1985. From that time to the present, I have continued to teach nutrition and the full spectrum RBTI. He left behind two daughters in Orlando as well as Laverne and Betty and a son Eugene, the latter three working and supporting him until his death. He also left a loving and beautiful wife, Bonnie, and an infant daughter, Stephanie, in Pennsylvania. He was 82 years old. Up until his death he had the skin of a baby. In reading the book you will find that Dr. Reams, while in military service, was blown up by a hand grenade. Some of the lead shrapnel, embedded in his body, could not be removed. His pancreas bore the insult of lead and in the fifth range of healing (0-20 [1], 20-40 [2], 40-60 [3], 60-80 [4], 80-100 [5]) of 80-100 years, his immune system could no longer fight back. Our gallant warrior for truth succumbed to cancer of the pancreas after a brief battle. I am grateful that he did not linger in pain. Funeral services were conducted in Pennsylvania before the body was shipped to his beloved Florida. Outside of Orlando on a sunny slope I gathered with his two Florida daughters, his brothers and their families and friends to say good-bye to a great man who had finished his job. Remembering there all the things he had taught me, and the thousands across the world whom he had taught the message of health, caused me to linger after others had left. As I held his hand, one last time I thanked him for touching my life. It is by the grace of God and the example and wisdom of this great man that I am alive today.

June M. Wiles, Ph.D., copyright owner
Holistic Labs, Inc., N.C.A., Inc.
Tampa, FL 33617

PREFACE

Holistic Healing and Free Radical Pathology as It Relates to Biological Ionization

(Balancing Body Chemistry)

Our nation at this time is focused on health care and preventive health care. It is overwhelming for all of us because health insurance is cost prohibitive: Doctor and hospital fees are no longer realistically priced; and the cost of drugs are out of sight! Americans cannot afford to become ill and yet they are, everyday! For many the cost of illness has become worse than the disease — with stress so great, death is a welcome release. This should not be so in a nation so bountifully blessed. What has happened to us? Where did we go wrong? Why did not we, the American people, pave the way to build for our own personal welfare, a more sure foundation in holistic health care? What can we do to correct it — is it too late?

After our forefathers, through strong spiritual guidance and determination for all freedoms, set in motion our great Constitution and Bill of Rights, America was now on course destined for greatness. Like spoiled children we never totally knew or felt or experienced our ancestors hardships of pain, hard labor or struggle for survival. It is as if comforts and plenty were handed to us on a silver platter. Two hundred years ago Dr. Benjamin Rush, one of the signers of the Declaration of Independence, warned that the American people should immediately develop a plan to keep its people free and flexible in matters of health care or eventually be gobbled up by a despotic medical machine.

Like we today, as time passed, comforts and ease of living in our great country came too easily and Dr. Rush's admonition went by the wayside. Before the American people had awakened, the medical drug cartel had woven its way into the fabric of our government, preventing and restricting our citizens in freedom of choice in health matters. A people who do not guard or protect its freedoms, like anything else, will lose them. Not only have we lost our pure air, the majority of our life-giving forests with pure medicinal plants and herbs and verdant lands, which are all being stripped away in the name of modern science changing many harmless plants into dangerous synthetic drugs, but also our clean water **and pure** foods, which are being handed back to us in chemical form called progress. Let me explain — with the advent of WWII, we spawned a whole new way of thinking. We accepted without reservations or knowledge of consequences the overuse of the **mid-century medical marvels in new drugs** and **food technology.** This lack of investigation has in many ways become a curse to us — we have become an **immunodeficient people, overusing what the medical profession has over promoted.** This age of chemicals has caught us even tighter in the vice of the despotic medical machine as well as the profiteering food purveyors. Immune deficient people can neither function intelligently nor physically well because of too many free radicals in their bodies. Shamefully this is where the American people are today.

What are free radicals and why are they so bad? Free radicals are microscopic forms of life that feed on decaying matter. Free radicals within themselves are not bad, they occur everywhere in nature — they are involved in the breakdown of dead things: Decay is a free radical **mediated process** returning matter to its simplest inorganic form. When we are cut or burned or exposed to toxic substances, stimulants, trauma or pressure or one or more of a thousand disease-causing bacteria, viruses, ameoba, fungi or any other form of parasite (and all are parasites if we play host to them) entering our bodies by way of eyes, ears, nose, mouth or other orifices or skin of the body, we begin to experience the affect of free radical takeover.

Another way we may suffer from free radicals is when the **immune system** becomes too weak to fight back from the above mentioned bacteria, viruses, etc., which may **already be in the body.** These free loaders may spring to life because the internal environment has digressed to the point that a suitable form of food (decay) is now available. A good analogy would be a dry desert with all types of seeds waiting for a good rain (a proper environment to bring forth life). Bear in mind that every disease in the body manifests itself with its own appropriate "parasite." At this point the medical practitioner may give it an appropriate name such as AIDS, pneumonia, strep, staph, ulcers, cancer, etc.

In the human body, cells of which we are made, are surrounded by electrons which further manifest themselves in neutrons and protons. These electrons will travel in a clockwise and counterclockwise direction around the cell, providing balance and life to the cell, much like our earth solar system (Warburg Theory), except, in the earth solar system the earth is spinning because of the Van Allen Belt and its solar system to provide the energy **force. The human cell remains intact with energy** being brought into the cell by way of the electrons — the miniature solar system. Of course, this is an oversimplification, but helps to explain.

When too many free radicals form because of a decaying environment, an acid condition (pH below 6.4) may develop — with this the electrons may spin too fast around the cell — so much so that energy may develop too fast and burn up healthy cells. This can also happen when too much sugar is eaten which provides food for parasites and promotes acidity. Also if the salt is too high (above 12) measured on the cellular conductance meter, then the salt becomes a scavenger, grabs protein and converts it to ureas. Urea — a waste product waiting its turn to leave the body, may combine with sugar and cause further fermentation, an activity of free radicals, thus more acid and a greater loss of healthy cells. If sugar is not a factor but salt is, the salt itself being an alkaline factor, may cause a "pickling syndrome" in the cell as well as in the colon and vascular system. In this instance, because there is no influence of sugar, the pH meter may register the urine pH as alkaline (7.0 and upward), not acid. Thus we see cells dying off because of the salt pickling syndrome and a pH too alkaline, another great environment for free radicals.

To clarify the subject of pH, the body functions best between 5.4 to 7.6 with 6.4 as ideal. Outside 5.4 acid and 7.6 alkaline the body is depleted of all minerals including hydrogen, oxygen, carbon and nitrogen. This information is gathered from the periodic table of atomic weights.

Free radicals can multiply very rapidly when the urine pH strays too far from the ideal of 6.4 pH — either too acid or too alkaline. The body strives at all times for a homeostatic balance of 6.4. There are in the human body approximately 10 quadrillion human cells and 100 quadrillion healthy bacterial cells called lactobacillus (a family grouping of approximately 35 and perhaps more) as well as the family of bifidus. These bacteria are in a ratio of (1) to (10) to keep the pH at a balance of 6.4 to feed us by further manufacturing from the food we take in more vitamins and minerals to keep us alive. A good example is the healthy bacteria in the alimentary canal that **prevent** gas, belching, burping and odorific stools. When these are unbalanced (away from 6.4 pH) from putrefactive bacteria, the gas and etc. just mentioned can be very uncomfortable and cause a premature loss of energy. Why do these putrefactive bacteria or free radicals grow so rapidly in the body? The electrons, an innate energy force in the body composed of neutrons and protons **are not alive,** but provide electrical energy to the body. Parasites **are alive** and are also energy. Electrons come in pairs and when hit or touched by another electron or parasite (an energy force), will lose its pair and immediately seek to pair up with another electron or energy force. Because many parasites (bacteria) are so small, in fact, the size of an electron, it/they may pair off individually with an electron. When this happens, putrefactive bacteria can multiply very rapidly. In fact, just one free radical paired with an electron can create millions or billions of free radicals (putrefactive bacteria) producing a rapid invasion of the body.

This is not to say that all parasites are microscopic — they are not. Some range from 25 feet (tapeworm), to hookworm, seven-inch worm, ringworm, pinworms, etc., right on down to the microscopic free loaders where they all eat your food, drink your blood, and leave their excrement in your body to be reabsorbed into your bloodstream as nourishment. If the immune system has been compromised in any way, parasites can, if left unchecked, take over and devour the body in a very short period of time. What does this mean to our immune system? It simply means that our ideal environment of a ratio of one to ten human cells to the inborn friendly bacteria cells has become overwhelmed with putrefactive bacteria or free radicals. Free radicals tear down the immune system. This is the state of disease. If this dangerous trend is not stopped by defender cells, such as leukocytes, lymphocytes, phagocytes and monocytes, etc., death can occur. When the body has reached this crisis point, we may have no other choice but to use antibiotics. Antibiotics and strong drugs may kill off the invaders but also kills off millions and billions of healthy cells, causing a gross imbalance in the one to ten ratio. In this state the immune system becomes weak, we age rapidly and the life span of expected 120 years (according to the book of Genesis), is cut in half. We're at the end of the line — how do we reverse the aging process? Obviously we must return to a holistic, preventive life-style. But what can we do about the free radicals — how can we stop the over-multiplication naturally? The answer: We must get the urine pH back to 6.4 where all nourishment and vitamin and minerals can be accepted by the body. How?

KNOW THE EQUATION — There is a perfect equation for every human being.

SUGAR	pH	SALT	ALBUMIN	UREAS
1.5-2.0	<u>6.4 UR</u>	6-7	Cell Debris	<u>3</u> N. Nitrogen
	6.4 SAL		.04M	3 A. Nitrogen

IDENTIFY YOUR OWN IMBALANCES — With this equation we can identify the location of greatest outlay of lost energy or diseased area. Naming the disease does not change the condition.

CHANGE THE LIFE-STYLE: I will elaborate on each of these.

(A) CHANGE THE DIET — Reintroduce healthy bacteria. Avoid foodless food and junk foods and get back to organic fruits and vegetables and grains without chemicals. Eat fresh scale fish without skin and shell, no pork (Lev, 11 Chap.), home fed fowl and beef. Take in more lactobacillus/bifidus and eat yogurt (plain).

(B) KILL OUT THE PARASITES WITH 6-N-1 — The colon complements (3) and antioxidants, and silvicidal, garlic, castor oil, K-min (primarily a special form of diatomaceous earth), walnut tincture, and soft rock phosphate. I am not here minimizing the importance of antioxidants. We have them in our line of products and we do know they will destroy free radicals, but I do believe of the same importance, we must get rid of the key reason (a dirty colon) for their existence.

(C) COLONICS, ENEMAS — To help rid the body of toxic waste (colon, lymph, and blood) — all are involved in the cleansing.

(D) GIVE CORRECT SUPPLEMENTS — To feed the body pure nourishment — the individual chemistry pattern (equation) will dictate this.

(E) APPROPRIATE AMOUNT OF WATER — Nothing happens in the body without water. It is nature's perfect solvent and required for every movement in the body. One-half the weight in ounces of water daily — never exceed 80 ounces — never drink with meals and for one hour past. Use only distilled water for drinking and cooking and only filtered or spring water for enemas and colonics.

(F) MORE REST — The body requires eight hours of sleep to restore energy. We must stay within the circadian rhythm of the 24-hour cycle. From

12 AM to 4 AM the body discards dead cells and brings in new healthy cells. From 4 AM to 6 AM the endocrines prepare the body for fight or flight of the day — i.e. — the trauma of the day. We should be asleep by 11 PM and never up before 6 AM, and 7 AM is better, to get the ideal eight hours of rest.

(G) MORE EXERCISE — Cleanses and reoxygenates blood. Muscle activity helps lactic acid convert to pyruvic acid to glycogen to glucose where it is recovered and stored in the fat cells and the waste product discarded by way of the urine. Active movement in the body prevents cross linking of tissue, assists in digestion of foods and evacuation of waste matter.

(H) FASTING — Discard diseased debris — use only distilled water. Seek guidance — important.

(I) MONITOR EQUATION — Until the person has learned to care for himself — that is, to get back to the ideal body pH of 6.4 by effective life-style changes, so **abnormal** free radicals do not form.

C.F.S. — An overload of free radicals in its purest form, this is Chronic Fatigue Syndrome. Who needs it when the energy can be completely restored without pain or drugs and returned to joy and vitality with a zip in your step and a song in your heart?

By changing the life-style you can avoid cutting, burning, radiation, drugs, toxic chemicals, trauma, too much stress and dangerous parasites.

(J) GET BACK TO THE PREORDAINED GOD DESIGN — And be obedient to the spiritual as well as the physical laws to free the heart and mind of guilt, anger, hate, envy, malice and other negative attitudes. When we overload the mind with these negative emotions, the death enzyme cathypain* takes over the physical and along with free radicals will return us to the earth. God wants His children and ambassadors for Him to be healthy, positive and enthusiastic. The God of us all wrote through the Apostle John, "I would that you prosper and be in good health, even as thy soul prospers."

The writer John put our prosperity, health and spiritual well-being on the same plane. By following the preordained plan for us, we can have the monetary comforts, abundant good health and a soul-satisfying coexistence with our fellow man and our God.

*Cathypain lowers the immune system (loss of resistance to disease). By nature, it is meant to move us out of the way to make room for those with more vitality.

NUTRITIONAL COUNSELORS OF AMERICA, INC.
1725 E. Fowler Avenue • Tampa, Florida 33612
June M. Wiles — President/CEO/Instructor — RBTI

PUBLISHER'S NOTE

Freedom is a word that has been sounded, spoken, sung from one end of America to the other for many years.

Freedom of religion.

Freedom of choice.

Freedom of the press.

Freedom to speak, or freedom to be silent.

Freedom to promote the progress of science.

America was founded on the principles that all men are created equal, and that we could each worship, serve and obey God as our will dictated.

Strange how this word called FREEDOM can become so confused with law, rules and regulations. Such is the case of Carey A. Reams.

I have spent hours, days, and even weeks discussing with Mr. Reams the health message of the Bible. I have watched him smile and weep as he described how God gave him the formula for perfect health. How God revealed to him in his laboratory the importance of obeying the Scriptures, even when it comes to the way we eat. He has suggested to people the world over the benefits of obeying the holy Scriptures. And for this he has been accused, persecuted, harassed, and yes—imprisoned. Not once, but many times.

Recently I found myself in Riverside, California. To my amazement this knowledgeable, sweet, gentle, spiritual, God-fearing old man had been placed in the Riverside County jail. The charge? Practicing medicine without a license.

No shots had been given, no drugs administered, no surgery performed, no cobalt, no chemotherapy, no radiation—just minerals, good food, pure water and vitamins.

America, when I realize that God is just, I tremble.

As I visited with him at the jail and went into the little cage and awaited Carey Reams' appearance on the other side I couldn't help asking myself: *Why, God? Why? Why would a man in his seventies, who loves You, loves people, and it is his desire that they be helped, suffer like this?*

I could hear the door creaking on the other side of the cage, and I looked through the streaked, tear-stained glass and there was "Doc." Our hands lifted to the glass, and our desire was to hold hands but it was impossible.

We looked at each other and our hearts joined by the Spirit of God.

Doc became very emotional as he saw me, and he kept saying, "Cliff. . . Cliff. . ." and held up two fingers. He did this several times without being able to continue his sentence. Finally, through his anguish and sobs he completed the sentence: "Two of my fellow inmates have found Jesus, and I'm working on two others." He told me not to worry, "it is all right, the Holy Spirit is in the cell with me."

Our time was soon up and we departed. His last words were: "Don't worry. No one can lock out from my cell the Spirit of the living Christ."

That was several weeks ago.

I have just received word while Carey Reams has been awaiting trial in the prison he became ill and has now been transferred to the Riverside General Hospital.

America, when I realize that God is just, I tremble.

—Cliff Dudley

ACKNOWLEDGMENTS

I wish to express appreciation to all those who have worked with me and stood by me in making this book a reality.

- — to my children, Laverne Reams, Eugene Reams, and Betty Brown, who worked diligently with me in my research laboratory from the time they were old enough to help until this day. Some have gone to prison with me and are also still willing to go to prison, if necessary.
- — to my wife for putting up with me all these years. It has been her prayers that have kept me on the battlefront for God. After one of our battles for the Lord she said, "Thank God He only made one like you!"
- — to Cliff Dudley, his wife and family.
- — to New Leaf Press, Inc., office staff.

* * * * *

This book is only a book of information for laymen, not intended for a text book for college graduate biologists, so therefore most college biologists and biophysicists are going to raise many questions that will not be answered in this book. Seminars are being taught to doctors, chiropractors, and people with degrees, those qualified with common sense to help this nation to be the healthiest on earth.

Chapter 1

EQUATION NUMBERS

One of the first things most people demand to know of a Reams tester is, "What is wrong with me?" This is a point that makes it possible for any of the Reams testers to be falsely accused of practicing medicine without a license.

There are biological laws in the Bible pertaining to health which teach us what is right and what is wrong. Unless the Reams Biological Theory of Ionization tests indicated what was wrong, we would not know what was right. It is absolutely impossible to separate religion and good health; or religion and poor health because we are all created by God.

Man's mind is in two divisions, one is the spiritual mind, or the subconscious mind, and the other is the conscious mind. With the conscious mind we think, we learn, we are taught principles of life, we are taught professional and non-professional duties. But back in the brain there is a subconscious mind, which is the spirit mind. This mind controls and maintains life. This is the involuntary part of the brain that keeps our organs functioning without any thought on our part of how and whether or not they function.

The Bible tells us that upon death the spirit shall return to God which made it and that the soul shall go down to the grave and return to dust from which it came. In this soul are contained the frequencies, the micronage, the milli-micronage and the milli-milli-micronage, and these contain our numbers. Here is one of the very important factors taught by science: pictures of the spirit leaving the body have been taken of people as they passed into eternity and it is recognizable. A camera can pick up the spirit as it leaves the body, ascending like a vapor or like a cloud upward into eternity, but the soul then goes back to dust, waiting the judgment day.

An undertaker who took the RBTI course said to me one day, after I had told him, "It's my duty to keep people out of your office and place of business as long as I can":

"That's why I'm here because I want them to stay out and live to a ripe old age." Then he added, "I was looking at bodies of adults in the morgue one day and this thought came to me: They are here because they wanted to be . . . a freedom of choice."

"What do you mean?" I asked.

"They broke every rule of health. They abused their bodies. They went to physicians instead of going to God for healing. They took drugs instead of minerals to get well. They did not seek to clean the temple of God. What other conclusion could I draw other than they were here because they wanted to be," he answered.

I'm not saying that I agree or disagree with his statement. I'm only saying that it lies within us to live to a ripe old age and to be healthy, or to be sick and die young.

I was in the South Pacific during the war and one of the things I learned about the natives there was that they never grow old. They all die young because their living standards are so low, their food is so poor and the quality of their life is thinking there is nothing really to live for. When you see one of those natives 30 years old, they look like 70 or 80. The conditions under which they live burns out their system too quickly.

This is what happens when the wrong foods are used. It releases too many calories of heat energy, or electrical energy at the wrong time and the body cannot assimilate it. The very fact that it does this, will show in the numbers, and disease will strike in the weakest place. Naturally the excessive heat energy will burn out the weakest place first, so the numbers in the test then come forth and we can zero in on the organ, or organs, that are affected.

All diseases start with one or more of the vital organs which are controlled by the central nervous system, chiefly the liver, but then spreads to other organs because the liver is the one organ that manufactures the skeleton of the amino acid for all the other organs, and the amino acid is the building block that we live on. The numbers help us to zero in on the location of the loss of energy.

A mineral deficiency is the cause of all diseases. The higher the specific gravity the less energy you may get out of your food. The lower the specific gravity, below the range A of the Reams Biological Theory of Ionization test (RBTI), the faster the food passes through the system and it does not have time to take the energy properly from the food. The numbers show what organs are losing the energy.

Collagen disease is nothing more than old-fashioned scurvy, caused by a lack of vitamin C, and affects the entire body chemistry from the top of the head to the bottom of the feet. The cells are coming apart, and the energy is being lost throughout the system. The beginning of this disease is the lack of calciums, which in turn causes a decrease in vitamin C, which in turn permits a weakening of the tissues, and a weakening of the gastric juices, which again affects the numbers.

The numbers indicate the state of health and how much *reserve energy* you have. The higher the reserve energy the greater liberty you have in breaking the rules. The lower your reserve energy, the less liberty you have in breaking the rules. You would not think of feeding an infant meat or nuts. Many people who are very ill, with a reserve energy rating below 15, their gastric juices are as diluted as an infant, and, therefore, they should live on baby food until their energy is restored, or increases.

There are two reasons why body chemistry will not respond to diet. One is a brain tumor in the left quadrant of the brain, and the other is unrepairable damage to the main vagus nerve (the vital nerve which carries the message from the brain) to the vital organs which supply the total daily nutrient (TDN) needed to maintain the maximum amount of reserve energy.

Suppose you were male, age 25, six feet tall, weight 180, and your numbers would read:

$$7.0 \quad\quad \frac{7.7}{7.8} \quad\quad 45C \quad\quad 4M \quad\quad \frac{14}{15}$$

The person with these numbers would be in a critical condition, and the body chemistry in range C. All of the above facts are necessary in order to interpret this equation because a baby six months old could also have these numbers, and therefore the diet would be different, or if a female had these numbers the diet would be entirely different.

The numbers indicate there are problems in certain areas of the body regardless of the age of an adult, but in a child under 12 years of age or under age of puberty, it would have an entirely different meaning because there is a change in body chemistry after a boy comes into young manhood, or a girl young womanhood, because of the different ratio of the calciums used by the female. Also, it's important to know if this is a caucasian, because the darker the color the more vitamin D the skin picks up from the sunlight and rays. The lighter the skin the less vitamin D the body picks up from the sun's rays.

The numbers indicate this male is a borderline diabetic. A man at 25 being a borderline diabetic means that the energy rating is going to drop down because if his pancreas manufactured too much insulin, or not enough insulin, then the body will not assimilate vitamin C. In taking insulin, the artificial insulin taken by tablets, or shots, will not make vitamin C available to the body, but the insulin that's manufactured by the pancreas will make vitamin C available. This condition might have just started a month or two before, or it might have started a few years before, but we have reason to believe by the process of deduction, that because of this carbohydrate condition the pancreas is not manufacturing enough insulin, and is of a rather short duration. The reason we know that is his weight is still 180 pounds. If he stays in this low insulin condition very long his weight will begin to decrease. Why would it decrease? The more sugar you have in the blood and in the system, the more nature will demand it and will begin to draw sugar from the body fats and carbohydrates from the muscles. It will leave a person hungry all the time. Their food is not satisfying them.

The second number $\frac{7.7}{7.8}$ (the urine reading is at the top, the saliva reading at the bottom) is the pH reading, which is a measure of resistance indicating that the food is digesting too slowly and that it lies in the stomach. Since the food digests too slowly, it is gradually creating pockets in the colon because the elimination is too slow. This number also indicates that the body contains too much calcium oxide and not enough calcium lactate and phosphated calcium, and also dolomite calcium, calcium carbonate, or tri-calcium phosphate. There is too much of one kind of calcium and not enough of five other kinds of calciums, which is creating a digestion problem because the liver has to have some of all six of the calciums every day in order to manufacture 5 to 6 billion enzymes necessary to keep the body in perfect health.

$\frac{7.7}{7.8}$ indicates that the liver is not manufacturing enough bile to properly digest the food, and practically no gelatin at all, which could prevent an easy elimination.

The pH reading is not a measure of the amount of acids or alkalines, it's a measure of resistance between acids and alkalines. By the resistance we can tell whether we have too much or too little calciums, or which calciums are enough and of which ones the man has too much. It is not a quantitative measure, it's a measure of resistance. By a differential calculus by deduction we can tell what is happening in the man's system.

The next number 45C shows the body is retaining too much salt and that the person is in the zone for a major angina heart attack, however, with his age it is probably 10 years away if this pattern continues.

Since the body is retaining too much salt it shows us how this reading is helping to exaggerate the conditions that we found in the pH $\frac{7.7}{7.8}$ reading because the food is too long in the colon, and now the salt is very high and is causing the intestines to lose their elasticity and stretch bigger and bigger, and as they do this forms pockets. If the ability to expand and contract continues in the colon it will get very much out of shape. The transverse colon will begin to sag, giving the man a "beer-bottle stomach" appearance, whether he drinks beer or not. (Drinking beer will only aggravate the condition.)

This high salt content indicates that cholesterol is beginning to form in the veins and arteries. If this person were to go to a doctor and have a blood test it may show that his cholesterol is normal because the blood changes every few minutes, and at the particular time that he was tested the cholesterol could have been low or it could have been medium or high. The RBTI test gives an overall picture of what is happening in the walls of the blood vessels, and in this person's case they are losing their ability to expand and contract.

The veins work differently from the arteries (I'll not go into that because it is rather technical), and unless nature puts oil in the veins and arteries to take the blood from the veins, from the capillaries back to the heart, the person will die much quicker because the vein would be so dilated until the blood cannot be pushed back to the heart. It is perfectly possible that his cholesterol, within a ten-year-period, would accumulate in such an amount that a piece of it could turn loose in the veins, go to the heart and block the heart and cause a very serious heart attack, or even death. However, this cholesterol would have to get above a saline reading of 47 total before it would be a fatal heart attack. Below 47 it would not be a fatal cholesterol heart attack. (i.e. there is no such thing as a minor angina heart attack. It is either Wham! you are dead, or you got lucky and lived.)

Now let's consider the last two numbers: $\frac{14}{15}$. Add these two numbers and you have the total amount of undigested proteins, which we call urea.

With the high urinary sugar, a high pH, and high salt, it indicates that this man is in a fatal heart attack zone and could have a fatal heart attack at any time. Anyone with this set of numbers should be very careful not to over exercise, to get too tired, too excited, lose his/her temper and "blow his/her top," or even get in a hurry because any of these things could bring about instant death.

Now let's look at the overall picture.

The numbers indicate that nature is trying to throw out the delta cells, but the man's diet is so narrow he is not getting the mineral content to maintain good health. He is not drinking enough distilled water. (He should divide his weight by two, which would be 90, and call it ounces—drink 84-90 ounces of water a day.) He should drink 4 ounces of distilled water every half hour for approximately 10½ or 11 hours a day, and then this will be the first step in bringing his numbers back to perfect. However, he could become discouraged. Anyone with this set of numbers trying to do this at home could cause the body to go into a withdrawal and cause a heart attack. Bringing these numbers from range C to range A in the RBTI test should be done under supervision where every number could be watched very closely. If the urinary sugar, the first number, drops below 5.5 the first day or the first hour, then the man could go into a violent withdrawal and go into a heart attack.

In the retreat we do not put the person on lemon juice and water immediately, but on water alone until the urea (that's the last numbers) is within a safe zone. After those numbers come down to a safe zone, and probably the urea down below 20 total, and the salt below 35, then it would be safe to put the man on lemonade (lemon with sweetener) and water to bring him quickly to between 2 and 5.5 on the urinary sugar.

In the meantime the salt would continue to drop and many times when the body chemistry starts to react it over-reacts and the sugar level would have dropped too low, too quickly and cause a seizure, or it could cause the person to go into a temporary coma, which is very easy to correct. The sugar level has dropped too low and there is not enough oxygen going to the brain. By putting a little sugar, or a little honey, on the tongue, in one minute he would come back to reality again, with no serious damage done. This is an ordinary occurrence in every retreat.

This man would only need to be in the retreat two weeks, but he still has serious problems. Whenever there's not enough insulin being manufactured to control the carbohydrates, there's not enough vitamin C for cuts, bruises or burns to heal. It's the beginning of serious problems ahead. Yet, if the average medical doctor examined this man with all the modern tests we have today, he would probably not find a thing in the world wrong with him, simply because the medical profession do not have tests that would indicate this problem. I'm not casting reflection upon anyone, I'm only stating a fact.

For example: Diabetes by the medical profession is determined strictly by the glucose in the blood. If there is more than 120 milligrams of glucose per gram of blood, the person is said to be a diabetic, and is generally put on orinase or diabinese until these numbers reach 145. After there is 145 milligrams of glucose per gram in the blood the person is generally put on insulin.

The medical colleges do not teach to compare a ratio

between the glucose and the total carbohydrates, which is very, very important. It is the missing link in the treatment of the malfunctioning pancreas. Many people become diabetic whenever the sugar level is only 80 milligrams of glucose per gram of blood. Others don't become diabetics until it becomes 150 or 160 milligrams of glucose per gram of blood. These facts are not considered by the drug establishment (manufacturers) as being important. What a shame!

One could have the sugar of 1.5 total carbohydrate in the urine and have a 300 milligram of glucose blood reading, or even 200 milligrams of glucose per gram of blood, and be perfectly safe and not be a diabetic. However, you would be a borderline diabetic, but it would not indicate that you needed insulin. Any doctor that found this would put the patient on insulin immediately. This is the rule they have to go by.

If you have a urinary sugar reading as this young man, of 7.0 and had a glucose reading of 80 milligrams or 100 milligrams of glucose per gram of blood, you would also be a borderline diabetic. But if you had the 7.0 and 140 milligrams of glucose per gram of blood, you would definitely be a diabetic. The point here is, rather than immediately putting someone on insulin, the first thing that should be done in order to try to prevent the use of drugs is to have this person drink 4 ounces of distilled water every ½ hour (until they drink ½ their weight converted to ounces in water) and then make another test and see whether or not the body responded. In this way most people can be kept off of insulin.

Insulin is a salt which causes the intestines to have diverticulitis areas (pockets), causes cholesterol to form in the blood, a breakdown of blood vessels, and causes hardening of the arteries of the brain. Insulin should never be given to children because the blood vessels are so delicate they will harden quickly, and most children that start on insulin die or become ill before they are 20 years old. There are some exceptions to this. Any of the RBTI testers can prevent your child from ever getting on insulin. It is unnecessary.

If your child should have cancer of the pancreas, insulin will not do any good, neither will diet. But cancer of the pancreas is so very rare. You might find one case in a million. If there is damage to the vagus nerve from the main branch that goes to the pancreas nothing can be done, and the body will not respond to diet, but if there is no damage to the vagus nerve, between the liver and the pancreas, the pancreas could quite likely get well even though it has been determined by medical doctors as being cancerous.

If you were told everything that could happen about numbers and the 2,600 differentials, if you read extremely rapidly, it would take you approximately 2,000 years to read if you read 12 hours every day. These are things we teach the testers if they take all nine seminars.

The above numbers indicate how marvelous a body God has given us. If all the water in all the oceans were converted to drops there would be only 1/3 enough drops to indicate what could happen to our body to keep it from being perfect.

Illness does not pounce upon us like an animal preying another animal, but illness is a result of deficiency of minerals in our body.

Chapter 2

DISCOVERY

The theory of ionization began long before this dispensation. It began with God.

Someone asked the question: "Where did God come from?" You will find that answer in the Book of Revelation: "Blessed is he that readeth, and they that hear the words of this prophecy, and keep those things that are written therein: for the time is at hand" (1:3). Then it goes on to tell what John saw and answers the question: "I am Alpha and Omega, the first and the last." Another translation says that God "has always been and always will be."

In the first chapter of the Book of Genesis we find God the creator and maker. After He made man He said, ". . . Be fruitful, and multiply, and replenish the earth." You cannot replenish something unless it had been inhabited before. Any geologist will tell you that there were at one time volcanos, volcanic ash, etc. So, God being a flaming fire, or flaming torch, evidently created things that were quite hot in the beginning, and therefore the heat in itself brought into being the process of ionization, of putting things together and taking them apart.

If you read farther in the Bible you will find that Lucifer, the archangel of all angels, made war in heaven and one-third of the angels made war with the other angels and God cast Lucifer down to the planet earth. And then, apparently Satan put his plan into existence and the waters became mixed with the dry land, and the sun refused to shine. Satan didn't pay his gas bill, or something of the kind because if the sun wasn't shining and the ionization of the spheres were not taking place, the earth would stand still.

The ionization of the spheres is the energy that God makes to hold the stars in place. The stars are suns, and around the stars the planets and moons rotate. Some planets have more than one moon, but all only have one sun. The closer you get to the sun the faster the planets rotate. The further away from the sun, the longer it takes for them to rotate. For instance, one of our years equals about three of Venus' years, and with Saturn it would take approximately 12 of our years to be one of its years. So the closer the sun is to the planet the faster the planets rotate—it is the increase of energy which is an increase of speed (the same as in an automobile, or anything that moves).

The greater the amount of energy, the faster the plant grows. Animals are not controlled by this same energy because they have glands. Then we can conclude that anything that doesn't have glands, like a plant, the process of osmosis is not limited by time, but animals that have glands are limited by time. A plant within a few weeks or months from the time it is seeded can reproduce itself, but people have to be in their teens before they are able to reproduce. Animals and birds have to be in different ages to reproduce. This is what we mean by the ionization of the spheres—the energy that makes anything click, or the energy that keeps it going. God in His great wisdom and mercy planned (Genesis 1) a new way to reestablish or replenish the earth.

In Genesis 1:1 you will find these words: "In the beginning God." He always has been and always will be. He created. This is God the Father who then formed the blueprint for this dispensation in which we find ourselves.

The first thing He made was LIGHT. And the sun began to shine again, and the earth began to rotate. The subterfuge of the earth turning caused the waters to separate from the dry land. Some people question whether or not God did this in a minute, or whether He took a thousand, or thousands of years to do it. As these first events were taking place there was no time. There was a rhythm but there was no time. Therefore, if there is no time, everything that was done had to be done instantaneously, and God spoke and it was so.

Whenever you begin to deal with ionization you are no longer dealing with geometrical math but relative math.

In relative math everything comes back to its starting point. A wheel turns. The earth rotates. Columbus proved that the earth was round. These are facts that are relevant.

In the field of geometric science there are straight lines and angles. In relative math there are no straight lines and no angles. For example: In geometric math the nearest distance between two points is a straight line. But in relative math this is not true, it is not the nearest distance. If we wanted to go straight from a certain place in the USA to Moscow, Russia, we would have to go straight through the earth for that is the nearest line, or the shortest distance. However, you can't get there that way so you have to go around the earth. This is not taking a straight line but an arc. That is relative math.

Another example of relative math is putting man on the moon. Man went from the USA to the moon, and in order to get back to the earth man did not turn around and come back to the earth, he continued in the same direction so that he could catch the earth in its orbit from the opposite side of the moon.

The mysteries of God are so great and so magnificent that it is very difficult for a human mind to conceive of the greatness of God. In the first and second chapters of the Book of Genesis is given the blueprint of this dispensation and the formation of our solar system. In the Gospel of St. John these words verify these facts: "The Word was made flesh." (This is speaking of Jesus Christ, the Messiah.) "In the beginning was the Word, and the Word was with God, and the Word was God." The word GOD means Father, Son and the Holy Spirit. It is the name of the deity. But there is a difference in God the Father, God the Son and God the Holy Spirit. There are denominations that teach that there is only one God. We agree. However, we agree it is God in three persons.

"The same was in the beginning with God. All things were made by him." This speaks of the Son, for it says, "In the beginning was the Word, and the Word was with God, and the Word was God. All things were made by Him, and without Him was not anything made that was made. In him was life; and the life was the light of men. And the light shineth in darkness; and the darkness comprehended it not" (John 1:2-5). There was darkness upon the face of the deep, and the sun began to shine. There was a shadow, and there was light. But where Christ is there is no shadow.

"There was a man sent from God, whose name was John. The same came for a witness, to bear witness of the Light, that all men through him might believe. He was not that light" (John the Baptist was not that light) "but was sent to bear witness of that Light. That was the true Light, which lighteth every man that cometh into the world. He was in the world," (this means Jesus, the Messiah) "and the world was made by him." This bears out the fact given in the second chapter of Genesis—the world was made by Him, and the world knew Him not. "He came unto his own, and his own received him not." This pinpoints for us that the Son of God, the Messiah, created or made everything for which the Father had made a blueprint.

How did He do it?

Fifty years ago we thought that nothing was smaller than an atom. Later we were to find that every atom was a little galaxie within itself.

To give an idea of how small an atom is: There are approximately 15 million atoms in one drop of water. An atom is a very, very small element, and is thought of as being mostly in elementary form. Once you leave an element then you have a compound which is composed of two or more elements, but a single atom must be in elementary form.

There are only 120 elements, and man has discovered approximately 110 of them. Some of the elements in their pure state would be: aluminum, carbon, gold, helium, hydrogen, iron, lithium, manganese, magnesium, nitrogen, oxygen, phosphorus, potash, potassium, silver. It would take a book to explain the ionization, or the way one of these elements came into existence.

If the atom isn't the smallest element, what then is the smallest element that God created? The answer is: ION. We say ION but actually an ION can be either an ANION or a CATION (pronounced, cat-eye-on). There is nothing smaller that God the Messiah created. After you leave that there is absolute nothingness.

What is a single anion? It is the smallest amount of energy in existence. The discoverer of this was Mr. Milhouse, and he called it a "Milhouse unit of energy." This term is no longer in use. It was a very common term fifty years ago. It means a millionth of a millionth of a millionth of a particle of energy, until there is no more. There is nothing else to divide.

The next smallest substance that God created was the CATION. The smallest cation contains 500% more energy than a single ANION. All anions were not singles but are *combinations* of anions. Therefore we must conclude that a single anion has a Milhouse unit of energy of from one Milhouse unit to a maximum of 499 Milhouse units of energy. If this then is the limit of the power of an anion, then a cation must also have its limits.

The limits of a cation is from 500 Milhouse units of energy to 999 Milhouse units of energy. If it took on one more Milhouse units of energy it would split into two and become two cations, or 500 Milhouse units each. We are now peeping into the mysteries of the creation of this entire universe. It was once said that "He who controls the atom can control the universe." That is true, but it is more accurate to say, HE WHO CONTROLS THE ANION AND THE CATION IS THE FOUNDER AND MAKER OF THE UNIVERSE.

Anions and cations form the elements into nine geometrical energy patterns. It is a mystery to us today how there are 120 elements and all of these elements and compounds are made with only nine different patterns of energy. In any element you can have a variable number of anions and a variable number of cations within the same element to make it form into one of these nine patterns, but the energy is a synchronized energy which equals the specific gravity.

When most people think of energy they only think of heat energy. But heat is only one kind of energy. When we think of heat we must think of patterns of energy. Think of two kinds of patterns: anionic and cationic patterns. The cationic heat has various shapes, and the anionic heat energy has but one shape. This is another mystery of God. All anionic energy is shaped exactly alike whether it is one Milhouse unit of energy or 499 units of energy. Cations have nine patterns of energy.

The interior of an atom is called nucleus, and the outer is called the shell. The entire nucleus of an atom we will call an ION, and it may be a cation or an anion, but never mixed. An electron may be a cation or an anion, *but an ion is always the interior of the shell,* and the electron is always the outer shell itself. This is the path that the electrons travel that form the nine patterns of energy.

When you stand in front of a heater, you are being bombarded with anions and cations of energy. They are so small, the anions can penetrate right through you without even touching you. It is like looking inside a box of grapefruit. You can drop marbles down between the grapefruit to the bottom of the box without even touching the grapefruit.

There are three kinds of energy.

Heat energy ($E_1 = MC^2$). In this book E_1 is heat energy.

Electrical energy ($E_2 = E_1 + MC^2$), which we will call electrical energy.

Matter ($M = E_1 + E_2C^2$). This too is energy which is solidified, as is found in rock, any vegetable or fruit. The greater the density, the harder and the heavier the weight of any matter whether biological or not.

This then lets us peek into the mysteries of the creation of this universe. It is called the process of ionization.

The process of ionization is LIKE THINGS ATTRACTING EACH OTHER. Nature follows a line of least resistance, and as it follows the line of least resistance like things attract each other. We have a proverb: "Birds of a feather flock together." So it is with the elements. Like elements attract each other. That is why gold is found in veins. Precious metals are found in deposits because these deposits attract from the entire earth the things that are like them.

Nearly 100 miles above the earth is the Van Allen (radiation) belt. This belt is the shell that completely surrounds this earth, and if it were not for that shell around this earth this planet wouldn't turn because the radiation of the sun strikes the Van Allen belt and creates a static field between the Van Allen belt which is anionic and the earth which is cationic and causes the earth to rotate, or any other planet to rotate. The moon doesn't rotate because it has no radiation belt. This is the law of physics that God has made. The same law which causes this earth to turn, causes an electric motor to run, or causes an engine to turn a dynamo to make or convert metallic energy into electrical energy. The same principle, only it is one principle in reverse of another principle.

Most people were taught the principles of chemistry which are not correct, and this is why so many answers are not found. They were taught incorrectly the basic rules by which God made this universe. For example, in the chemistry books we were taught that an atom has to have the same number of protons as neutrons, or, the same number of anions as cations. This is not true. If it were true then all iron would be exactly alike. All fruit would be exactly alike. There would be only one kind of grass. There would be only one variety of trees, and all human beings would look exactly alike. The truth is that all atoms under the same temperature and pressure are the same size but they do not weigh the same. (I realize that this statement will create much controversy and some will label this author a quack, but as they begin to prove these statements they will find that these statements are true.)

We have said that like things attract each other, and nature follows the line of least resistance. Since nature follows the line of least resistance then any atom with the same number of protons and neutrons or anion and cations, would be a hydrogen atom, with the ratio of one to one. Already we have conflict with the logic of God with what is taught in the chemistry books. We have two statements that are just the opposite.

This process can be more easily understood in considering compounds, such as H_2O. H_2O means 2 atoms of hydrogen and 1 atom of oxygen combined to form one molecule of water.

H has 1 anion and 1 cation

O has 1 anion and 16 cations

If anions can be from 1 to 499, and we choose to assign the smallest amount of power (i.e. 1 Milhouse unit of energy) for the anions; and if the cations can be 500 to 599 Milhouse units of energy and we choose to assign the smallest amount of power (i.e. 500 Milhouse units) for the cations, the total amount of energy is just 1 molecule of very dry, light water = 9003 Milhouse units of energy:

	Anions		Cations		
H =	1	+	500	=	501
H =	1	+	500	=	501
O =	1	+	16×500	=	8001
					9003 = water (H_2O).

H_2O is two parts of hydrogen and one part of oxygen. Hydrogen has one anion and one cation in its joining to another molecule of two of the hydrogen atoms—which would be two anions and two cations as joining to another molecule of oxygen that does not have 8 electrons (or anions) and 8 protons (or cations), but does have one anion and 16 cations as oxygen. So we have water: H_2O.

There is no way that energy can be calculated or figured by mathematics with the old theory of equal number of anions and cations in an element, unless you figure the cationic energy versus the anionic resistance. There is no way to figure energy, because the total number of Milhouse units determines the total amount of energy. It is not possible for anything to give up this total amount of energy instantaneously without creating an explosion. In other words, the anions and cations would fly apart like a covey of quails that all flew away at once. In the natural decay anions are given up rather slowly, and the cations are given up very slowly.

This isn't the realization of how things are put together, but how things are taken apart. This is done by noting the structure of the number of anions and cations that make up the compound, element, or metal. For instance, there are many different kinds of pure iron because of the different number of cations and anions that compose the pure iron.

There are many different kinds of water, which is a compound, whereas iron is an element. There is heavy water. There is light water. There is just ordinary water, and then there is dry water. The less the number of Milhouse units in the water (I'm not speaking of impurities in the water), the lighter the water.

The greater the number of anion units in the Milhouse units in the nuclear atom and the higher the Milhouse units in the cation or electron in both the hydrogen and oxygen, the heavier the water. The greater the gravity, hence the difference in pure water. In this calculation of energy, we use 250 Milhouse units of anionic energy and 750 Milhouse units for cationic energy averages in working out the components of energy in the component total amount of energy in any given substance.

So, then we have now found our basic foundation for organic and inorganic chemistry:

ORGANIC chemistry is any substance containing carbon.
INORGANIC chemistry is any substance not containing carbon.

Another field of chemistry that is seldom mentioned, is the field of COLLOIDAL chemistry. There are two meanings or definitions of colloids. One is a measurement of size—*ohms*. In this field you may have colloidal elements like colloidal gold, colloidal iron, colloidal silver, etc., which is strictly measuremental size, only very, very small particles (like dust or fuller earth). In the other field the colloids are compounds.

In the field of colloidal compounds each atom is a complete little solar system within itself. It has its sun and its moons and planets rotating around it. There are many different numbers of moons and stars rotating around the ION CORE of a colloidal compound molecule. This is very important in the formation of matter.

Colloids are small, very much like the atoms, except a colloid is the smallest amount of an element that can exist, and the colloidal part of that matter is the smallest amount of matter that can exist in either elementary or compound form. The compound colloids are so fine until one cubic inch of it would cover 7½ acres with a solid sheet of colloids. They are also a very fine, sticky, glue-like substance when they come in contact with moisture. But where there is no moisture, colloids will stand in the air as in a vacuum because they are so small until even gravity does not attract them, until they collect enough moisture to weight them down to the earth. Did you ever notice a beam of sunlight shining down through a dark room, and the solid particles (very fine round particles) floating through the air? Those are a *group* of colloids because you cannot see one *single* colloid.

Colloids are very important in nature. The colloid in a tree makes it possible for that tree to stand erect, to go upward or perpendicular to the earth. But a vine does not have enough colloids in it for it to stand up, so it has to find something to climb on, and it climbs on something that has enough colloids to hold itself up.

Without colloids we wouldn't have any fingernails, toenails, hair, bones, or any substance to our system. Our bones and our teeth contain more colloids than any other part of our body because they are harder. Our fingernails and toenails contain the second greatest, hardest amount of colloids.

Did you ever walk out in the morning and see the "diamond" in the dew drops? This "diamond" was a carbon colloid.

Without colloids there would be no life upon this earth.

We have mentioned that above the earth is the Van Allen belt, and in order to understand why trees grow perpendicular to the earth (and not lying on the ground like a vine), we must understand that the earth is cationic and the anions that are given off from the cationic structure of the wood or bark of the tree is greater than the cations which attract it to the earth, or hold it in the earth, and this causes the tree to reach outward toward the Van Allen belt.

In a man's brain there is more potassium than there is in the feet. Potassium is an anionic element and therefore the brain is in the head, and we walk on our feet instead of our head. If our feet contained more potassium than our head then we would walk on our head with our feet in the air.

Animals have a greater cationic structure in their system than anionic and therefore they walk on all fours. The animal that has to depend on smell more than eyesight has to turn his nose nearest the earth because his sensation of smell leads him to the environment that he desires. When he is erect he holds high his head to see and smell better.

Soft iron contains a very high carbon content. By heating the iron and cooling it, it becomes harder as the carbon content decreases. This is taking the carbons out of iron and the iron becomes more pure. But the impurity of iron softens it. Practically in all substances—the more impurities that are in that substance generally the softer it becomes. I'm speaking of most of the carbons. The more carbon that is in the substance the softer the substance becomes, with the exception of the diamond. However, this is only true in organic chemistry, and may be true in some cases of inorganic chemistry.

All animal matter and plant matter contains carbon, and therefore is organic. It does not necessarily have to be decaying leaves or manure.

Many people think that organic means that it comes from

plant or animal matter. As far as the chemistry books, and as far as labels on goods in the store being organic or inorganic, it means that it has carbon in it. Honey has carbon and therefore is organic. One time I saw in a health food store a sign: "Organic Honey." I told the clerk I didn't care for the "organic" honey, and ordered five pounds of inorganic honey.

"Sir," she said, "I don't think the owner will buy that kind, we only sell pure products!"

You cannot have sugars without carbons, it is impossible ($C_6H_{22}O_{11}$ is one carbohydrate). The C stands for carbon, and you cannot have a carbohydrate without carbon. I was telling my wife about this and she said, "I'll be your inorganic honey!"

Chapter 3

CREATION

It is rather amusing when we begin to study nature. When we really begin to know nature we see not only trees, plants and people, but we see a frequency. Physicists, who are trained in this field, have to see many things. Like a field of orange trees, a field of peach trees, a field of pecan trees, or a church house full of people, or a city full of people. We see things that are attracting each other, and people only see the bigger solids—as a person—this is a result of a group of solids which makes us what we are.

To the physicist it is mathematical numbers, and to God who is a God of mathematics, He says even the hairs of our head are numbered. Well, if Jesus had said, "When your atoms and amino acid molecules are numbered" the people would not know what He was talking about. He knows not only what your atoms are, the numbers of your atoms that you are made of, but He also knows the number of your anions and cations. Every tree, every drop of water in the ocean (no two drops are alike), every snowflake (no two are alike), every blade of grass (no two are alike), reveals God in His greatness. When we understand the mathematic and chemical structure of matter we see how truly great God is, and how little we are. How insignificant we are, and how helpless we are, and how much we need Him to help us understand these mysteries.

The study of nature is so beautiful, so fantastic that only God can call it into existence.

In the Book of Amos, in the Old Testament, Chapter 3:7 we read: "Surely the Lord God will do nothing, but he revealeth his secret unto his servants the prophets." And in Jeremiah 33:2-3 it says, "Thus saith the Lord the maker thereof" (the Lord is the

Messiah) "the Lord that formed it, to establish it; the Lord is his name. Call unto me and I will answer thee, and show thee great and mighty things, which thou knowest not." God has kept His Word.

I trained as a pharmacist when I was in premed school. When I want to know something I ask the Lord. Unless I had had the training that man gives I would not have known what the Lord gave me in the Biological Theory of Ionization. It is like the little girl who asked her mother: "Where do babies come from?"

The mother answered: "Daddy plants a seed, the mother takes care of it, and then a baby comes."

A few days later the girl was playing out in the yard and found a big peach seed. She said, "I'm going to plant this seed and have me a baby." She planted the seed, and found a broken soup bowl and turned it over the seed. Two days later while playing in the yard she said, "Oh, I'm going to see if I have me a baby." She turned the bowl over and there sat an old toad frog. She said, "You ugly, old toad, you. I would kill you if I wasn't your mommy."

Even so today we do not understand these mysteries many times in what God says, and we have to be trained and become mature enough to understand the mysteries.

One of the great difficulties is that when we do understand the mysteries we refuse to believe them. In the greatness of God we are continually trying to bring God down to our size. Hence these mysteries are forever hidden from us. Also, we ask God for things in our prayers, and then when He sends them to us we refuse to accept the answer the way he sends it. Henry Ward Beecher said that when a housewife prays for patience the Lord sends a contrary cook. So when you ask God to reveal to you some of His mysteries, He might make you go back to school. You may say, "Oh, I don't want to do that. It takes too long." But God is not in any hurry. His days are not numbered. It always has been and always will be thus. We can accomplish more by preparing ourselves well for the task that God gives us, rather than going out ill-prepared, not understanding His great mysteries.

The earth changed direction after the flood. It rained 40 days and 40 nights. If 25 inches of water had fallen per hour, which is unheard of on the earth, there would have been less than 2,000 feet of water. Many mountains are five miles high, and the Scripture tells us *water covered the earth*. The Bible says there was a deluge of water. So if the earth stopped turning the water would cover the earth with equal depth all over. The Bible causes me to believe that after the flood the earth began to turn in a different direction.

From the fall of man till the flood it had not rained, and yet there was Noah building a boat out of gopher wood on dry land. He preached for 100 years and didn't get one convert (apart from his own family). Up to that time there was not a rainbow. If it had rained there would have been a rainbow with the sun shining through it. The Word tells us that the earth was watered by the dew and not rain. If the earth was turning in the opposite direction it would be anionic and there couldn't be any rain. But if it turned in an opposite direction it would be cationic and there would be rain.

There is no such thing as anionic water. There couldn't be any clouds to cause rain because it can only become rain when it is heated to a vapor. It is still cationic but the anions are in faster motion and the water is giving up cations so rapidly until the moisture rises—now that's pretty dry water in the clouds—it is steam or vapor.

Before the flood when the earth turned in an opposite direction, the Van Allen Belt, which is about 100 miles above the earth today, reached out nearly right to the earth, so any vapor that arose would be disintegrated into anions and cations and could not form clouds. Now, when the earth turned in the opposite direction, this then moved the Van Allen Belt out to where it is at the present time because of the difference in a cation and an anion, or a negative and a positive, which are in the direction in which the charge travels. So if the earth turned in an opposite direction then the earth would be cationic, energy would be coming out from the earth, flowing away from the earth instead of the earth turning in the opposite direction drawing the anionic belt closer to the earth. After the earth began to turn in an opposite direction and the deluge was caused because the earth stopped turning, then the vapor from the waters went up into the sky and formed clouds of various kinds. When these clouds come into contact with certain waves of cold air and a sudden change in temperature you will find that it causes condensation which is rain. Even thunder is caused because hot air is striking cold air. When this happens quickly it also sets off an electrical charge of electricity between the anions and the cations.

What causes the clouds to drift low on a certain level in the sky?

When the atmospheric pressure becomes greater because of the cold air then the steam, or water vapor of the hot air, causes the ceiling of the cloud to travel lower through the atmosphere. When the atmospheric pressure becomes irregular and it is very rough this causes the clouds to be cumulus, and as the steam then is condensed by the pressure of the colder air, that is the rain that falls. When the air is so condensed just ahead of the condensation there is a conglomeration of the anionic forces, and the cold air forces the anionic static electricity in the hot air rising, until a short circuit forms because of the magnetic field that is set between the earth and the cloud level in which the anions and cations touch each other. It creates energy in chain reaction, comes in contact with an anionic chain reaction, which causes a bolt of lightning to strike, generally it comes near a tree, steeple, building, and the nearer the earth the electrical charge gets, the greater the number of cations it picks up. The heat lightning is static electricity.

When you see heat lightning it is picking up more anions and thus stays up in the atmosphere. When it picks up enough cations then the charge comes toward the earth. I have been on the mountains when the static field would become so great your hair would almost stand on end and the hair on your arms would tingle from the static electricity. I have seen the lightning strike, and it looked like the electrical force came out of the earth and went up into the sky rather than coming out of the sky and going into the earth. It is a phenomenon that no one can deny because you see it right before your eyes.

Therefore this theory of ionization, of the story of creation, is becoming a reality because God is the God of math, a God of science, and He made it. He is a God of fire brighter than the sun. Fire is an anionic substance. This theory of ionization is helping us to understand some of His mysteries that have been of long-standing.

Chapter 4

HOMO SAPIENS

Since it is Homo sapiens, or human beings, we are involved with, we are going to stick pretty much to this throughout the rest of the book.

We will discuss the Reams Biological Theory of Ionization (RBTI).

The reason that I say the Reams Biological Theory of Ionization is that there are now a number of counterfeit theories of ionization, which is not the one that God gave me in 1931. I would have no objection to anyone using the Reams Theory of Ionization if it accomplished the purpose of the goal, as was mentioned in the Preface of this book. Some people are so afraid that they might be called upon to pay a royalty if they used the perfect equation, and therefore have counterfeited one of their own. I would be delighted to have them use the perfect equation rather than the counterfeit, because I want America to be the healthiest nation on earth. Provided they conformed to our laboratory guidelines that God has given through years of research.

In order to understand the difference between a cation and an anion is like the understanding between a negative charge and a positive charge. The difference in a negative charge and a positive charge is this: *it is simply the direction in which the charge or electrical current is traveling.*

In a substance wherein the cation is traveling counterclockwise in each nuclear shell, the substance is cationic. In an anionic substance the electron is an anion traveling clockwise in each molecular shell. It is something like the clock hands going around when you face the clock, the hands would be going clockwise—

which is anionic. But if the hands traveled in the opposite direction they would be cationic. The electrons in an atom travel in such a way that they form a complete shell around the atom like a mantle. This makes a world of difference in molecular structure. Because of the difference in the anionic and the cationic molecular chemical and mathematical structure of the micronage, the milli-micronage, and milli-milli-micronage, their synchronization divinely and physically determines the frequency. Now the frequency also determines the shape and form of all kinds as well as man, so we are truly marvelously and divinely made.

A tree stands straight up, and a vine has to climb a tree, but when a vine climbs a tree whether it is south of the equator or north of the equator it still climbs that tree counterclockwise. It goes around from the left towards you to the right—around and around that tree or post. It is climbing the tree counterclockwise, but the root (if it were in a pot) and you needed to replant the vine because it had become rootbound in the pot, you would find that the roots go around the pot clockwise.

Has it ever crossed your mind why that above the ground the vine climbs the tree or the stake counterclockwise, and yet the root goes around the pot clockwise? Actually it is the same direction. The only confusion is that we start in the middle. If you begin at the bottom it is the same direction all the way to the top. If you start at the top and go to the bottom it is the same direction all the way to the bottom (e.g., looking at a coil spring on end). God is not the author of confusion, He is the author of harmony.

What in the world does the direction of the electron and the anions and cations have to do with diet or health?

Unless we understand how one cell is made in our bodies we cannot understand how two cells, four cells, eight cells, and sixteen cells and all of the cells are made. In order to understand what manner of micronage we have, we must understand how minerals are made available and how to replace worn-out cells.

This brings us to know that we do not live off the food we eat, **WE LIVE OFF THE ENERGY IN THE FOOD WE EAT.**

What kind of energies are there?

How does one express those energies?

Einstein expressed heat energy by saying E equals MC

square ($E_1 = MC^2$). To most people who are not mathematicians that doesn't mean anything, and to many who are mathematicians it doesn't mean very much. In simple language we could say: **by burning matter, heat and electricity are formed.** Heat is electrical energy, which would be: E_2 equals the substance between E_1 and MC^2. Then there is matter. Matter is the substance composed in the union of $E_1 + E_2C^2$ = substance.

E_1 is heat
$E_1 = MC^2$
E_2 is electricity
M is matter
C is conversion

The square (2) is the process of ionization which puts things together or takes them apart.

$E_2 = E_1 + MC^2$
$M = E_1 + E_2C^2$

Elements and compounds are made from anions and cations. Compounds are made from elements. We now have the basic foundation for expressing energy in mathematical terms. However, this is not the whole picture. It is only a part of it because the resistance between anions of different values will give off heat energy. And the same is true of cations—those of different values when mixed together will give off heat energy.

We can say then that energy is made from the resistance between anions and anions, cations and cations, but the greatest amount of energy is made from anions and cations all seeking their synchronization point.

Nature makes energy in many compound patterns. And it makes these compound patterns from nine elementary geometric patterns.

The liver manufactures a substance called bile, which in its purest state is a form of dilute hydrochloric acid. The body stores bile in little sacs in the stomach, gall bladder and capillaries of the liver.

All the foods that we eat are cationic, with the exception of lemons. Lemon juice is anionic. In other words, the electron in lemon juice is traveling clockwise and all the electrons in other foods are traveling counterclockwise. However, some foods are more cationic than others, and some are less cationic. But we can say that lemon juice is nature's own natural dilute hydrochloric acid, and the liver can daily convert it into some 5 to 6 billion different enzymes, or vitamins if one is in good health. Lemon juice has less chemical change than any other natural substance known to man. The only other substance that can be used is man-made hydrochloric acid tablets.

It is rather strange too that we call hydrochloric acid an acid when it really isn't an acid, it is a base. In bases the electrons are anionic, therefore they travel clockwise in the molecule, and acids are cationic and therefore travel counterclockwise. So this is the actual physical difference between an acid and a base.

The foods that go into our stomach are cationic, when the bile is released it gives off heat and electrical energy in both anionic and cationic form. The energy from the digested food is taken to the liver, where it is mixed with oxygen, and also the natural calciums and many other elements in our diet.

The enzymes and the vitamins are the same thing. There is no difference. The liver joins these enzymes together in the skeletal forms of amino acids. It is not a complete, finished amino acid. When we use the word *complete* we think of it as perfect—but this is *unfinished*, it is in skeletal form of amino acid. The liver dumps this unfinished amino acid into the blood stream. The blood stream takes it by some 284 transformer glands in our body. It then either adds anionic or cationic energy to the amino acid, or gives off something through the glands, according to the magnetic micronage structure attraction of the glands. As the blood carries the amino acids through the body they become the building blocks for our system. Each organ takes the kind of building block from the amino acid in cationic form that it needs to rebuild and restore and keep perfect that organ.

This cationic energy is then a form of energy in one or more of the nine elementary, geometrical patterns. This is nature's way of putting together the various organs. The bones differ from the fingernails, the brain differs from the lungs, and the lungs differ from the liver, and the liver differs from the kidneys, etc. All are different structures but have the same frequency.

This brings us now to the point of molecular structure. How is one molecule of any organ made? We will begin with a person from the time they are born until they die.

You may have been taught that cells divide—which is incorrect. Cells do not divide. If cells divided they would then break the link of the nerve from the main source of the brain to the cell itself, and therefore it could not function. It is in the structure of the brain that the molecular pattern of the organ is restored, and also the anionic structure of the cell is formed, and also in the anionic and cationic structure that forms the atom. This then gives us the basis of the structure of one cell which can be expressed mathematically.

What is there now to keep a cell from just growing and growing all out of proportion?

This takes us back then to the brain. The brain is the determining factor of the magnitude and size of the cell. Man was made in the image of God. So if we are created in the image of God then we look like Him. Very much like Him. The cleaner we keep our temple the more like Him we become. Since that is true then there is something in the mathematics of our body structure that makes us like Him. The question now is: What is that factor?

That factor is the FREQUENCY upon which we live. You might say wave length, or vibrations, which mean the same thing. However, frequency is the correct word. When God made man, Adam, He made him in His own image and Adam had a log frequency of 24 (.0000024). Then God made woman, and He made her on a log frequency of 26 (.0000026). Everything He created He created on a frequency according to his kind. He created a dog on a log frequency of 38 (.000038) for the male, and 40 (.000040) for the female. He created horses on a log frequency of 44 (.000044) for the stud, and 46 (.000046) for the female or mare. When He created citrus He created a different log frequency, an odd number with the frequency of 9 (.0009).

Animals and plants that have an even number for their frequency, require male and female. They have to breed, or, there has to be a cross between the male and female plant in order for fruit blossoms to form. In plums there are male and female trees. The sour plums are the male trees and the female are the sweet plums. It is the same with cherries. It takes a male and a female plant to make the cherry blossoms fertile, or to pollinate the cherry blossoms so they will be sweet or sour cherries.

In citrus there is an amazing fact that does not have any particular bearing upon the frequency but it just happens to be peculiar. The female blossoms have five petals, and the male blossoms have four petals. The female blossoms come out first, and just when they are ready to fall then the male blossoms come out, and then the bees take the pollen from the four-petal blossoms to the five-petal blossoms and fertilize it, then fruit sets on the tree. The thing that is amazing about this is that five plus four is nine—which just happens to be a coincidence. I don't think it has any real scientific value, but it brings us back to the first chapter of Genesis when God said: Go and produce after his own kind. The "kind" is cattle, moose, elk, deer, buffalo—all having the same frequency. There are many different species of sheep, and they have the same frequency. There are many different species of cats, house cat, bob cat, etc., and they have the same frequency. This is true of lions, squirrels, etc., which are in the cat family.

All plants, according to their kind have the same frequency. There are many different varieties of peaches, yet all have the same frequency. There are many different varieties of oranges, grapefruit, tangerines and yet they have the same log frequency. It is beautiful how God created each thing after its kind, and some of the species are so far off from the original and yet it will not cross with its own kind. But still the way God made or designed is shown in the Book of Genesis.

You might ask, "Well, if everybody is on the same frequency, why don't all people look alike? Why are there different species?"

There are different species because there are different ways to place the atoms together, in which the patterns of anions and cations even within the same organs come together to make what we call MICRONAGE.

The difference in micronage is the difference in species and organs within any individual. It is also what keeps one human being from looking exactly like another human being. It is the way the anions and cations are stacked together in the element which makes the compound, which makes the micron of the structure to the person and still be in the same frequency, according to its kind.

You might also ask, "If that is true, then why the difference in color?"

Milli-micronage is the path of the electron in orbit which determines the *color*. The light strikes the electron shell and a prism of light forms what is known as color. Where there is no light there is no color. Under milli-micronage we have milli-milli-micronage, which is identity. Even identical twins are different. They are not exactly alike in God's sight. No two blades of grass, drops of water, or snowflakes are alike because of the difference in the anionic and cationic molecular structure from which the atoms are made.

Every kind and species of life can be expressed in mathematical terms because of the above facts. The frequency, the micronage, make it possible for every degree of life from perfect health to bad health to be expressed in mathematical terms. The following facts prove that this RBTI equation is accurate, and without these facts there would be no way to prove that the equation of perfect health would be accurate.

It is the varying patterns of compound energy that determines the structure of plants and animals, the difference in various organs of the plant and animal.

This brings us to one of the essential rules of physics and all biological life: *Anionic plant food makes growth, and cationic plant food makes fruit.* There is no exception to this rule.

Plants that come to bearing age and shed their blossoms instead of making fruit have plant food that they are taking from the soil that is anionic plant food, or alkaline. If it is very acid (cationic) too early then the plant tries to bear fruit when it is too small. Many times women when they are pregnant crave sour pickles because their body is too alkaline, and they need more energy. By taking in something on the acid side (acetic acid or vinegar), it gives them more energy from their food, and therefore supplies the body with the energy needed to produce another person. If the body is too acid the mother does not crave pickles but is in need of more calciums in her diet. RBTI laboratory tests indicate the balance that determines resistance, which indicates cause and effect, which also may be used to calculate the potassium or the calciums, or any other element needed to cause the body chemistry to be perfect again. Nature does its best to do this without outside help, but the Bible says that in the time of the end "the earth shall wax old like a garment" (Isa. 51:6). This means that the soil will be depleted of many of its vital mineral elements, and especially calciums, which are so essential to human life.

All plants and all animals use more calciums by weight and by volume than all the other elements combined. It is a very necessary element in the human diet.

Calcium is a single and plural word. There are more than a quarter of a million different kinds of calciums. In this book we will use the word calciums to emphasize the plural aspect.

Calciums can be divided into seven different classifications. Six of these CALCIUMS are essential to biological life. There is only one of the classification of calciums that is fatal to biological life, which would be pure calcium hydroxide. However, it is very easy to add too much of one classification kind and not enough of the other five kinds, or any combination between these extremes.

The second element our bodies need most is PHOSPHATES. Without phosphates you will not have any bones, fingernails or teeth. It is the phosphate of calciums that form the foundation structure for our bones and teeth.

The next element in volume that is needed is POTASSIUM. I'm speaking of it in this order because I do not count water as an element but as a compound.

WATER is the best and finest catalyst known throughout all nature. A catalyst is something that joins two or more elements or compounds together without becoming a part of the union as such. As a minister may marry a man and a woman without becoming a part of the union. The minister would be the catalyst. Also, as a wrench would join a nut and a bolt together without becoming a part of the union. The wrench would be the catalyst, it is the joining agent.

Water is also the best of all cleansing agents known to man. There is nothing as good to cleanse this temple that God has given us as clean, clear, distilled drinking water. Most people do not drink enough water. I often have people say to me, "Mr.

Reams, I never drink any water."

My reply is: "What do you think you are? A Volkswagen!"

We are water-cooled mechanisms, and water is just as essential to us as it is to plants. If we do not drink enough water to wash out the worn-out cells then our bodies will temporarily store or collect these worn-out cells. Just as a bad apple in a barrel of good apples, it will make the other apples bad more quickly. This also happens in our bodies.

In our bodies these worn-out cells collect in the weakest part, often causing pain in that part of the body—the knees, or joints, or muscles. The doctor will diagnose the pain and treat the hurts, without knowing where those worn-out cells originated. However, he is doctoring the effect and not the cause. In the RBTI we work on the CAUSE and pay very little attention to the effect. The effect is the result of a cause, and the cause is a result of a mineral deficiency. Therefore, we can conclude that all diseases are the result of a mineral deficiency which may be expressed in a mathematical equation.

Chapter 5

SOME OF THE CHEMISTRY OF FOODS

Each food has its frequency number, according to its kind, the same as people have their frequency numbers. Once you know the frequency then you know the proper diet of all plants and animals. A horse can live on grass all of its life after it is weaned. Man's frequency is the highest decimal and as such requires the most complex diet. Frequency is how much time it takes for one electron to make one complete revolution around one molecule. This is measured in micro numbers (e.g. man's frequency log of 24, or .0000024 of a second).

The greater variety of foods that we can eat, the safer it is to maintain good health. Our foods are depleted of minerals today, and by having a great variety we come nearer fulfilling the mineral needs of our deficiencies. We lose *reserve energy* by limiting our choice of food intake.

In order to understand food chemistry we must understand that there is a great similarity in amino acids in animals and plants. There is a great similarity in every frequency, and the micronage, milli-micronage and milli-milli-micronage structure. That is, the biological physical mathematical principles are very much alike in both animals and plants. No amino acids can be formed without **FIRST** having an atom of nitrogen. It is the first requirement in the formation of any amino acid cell. Amino acid in its true form may or may not be a complete cell. In its perfected form it could be a complete cell because all biological cells are amino acid cells, but seldom do you find one that is complete. However, it is possible for it to be complete.

The **SECOND** requirement after the first atom of nitrogen (protein) to form the amino acid cell is calcium, or one or more of the groups of calciums.

The **THIRD** requirement to form the amino acid cell is oxygen.

The **FOURTH** requirement is phosphate, and the **FIFTH** is potassium.

In this manner all biological structure is made, plants using carbon dioxide and animals oxygen.

The magnetism of the union of these and many other elements on any given frequency determine the organ structure, regardless of the various frequencies.

Phosphates in plants have a different chemical action from what phosphates do in people. In plants the colloidal particles make cellulose, and in animals they are the bone structure. The greater the amount of cellulose in the plant the more tree-like the plant. The less cellulose it contains the more vine-like the plant.

The amino acids in plants are formed by ionization exactly as in animals. Plants do not have blood like we have, they have what we call sap. The sap rises because the amino acid is in anionic form. The sap goes up into the branches and the leaves through the process of photosynthesis. It uses the phosphate to join carbon, hydrogen and oxygen. The plants breathe in the carbon from the air and breathe out the oxygen. Carbohydrates ($C_6 H_{22} O_{11}$) in plants are formed in very simple terms as follows:

Carbon + water *(Phosphates is the catalyst)* = sugar in plant

6 carbons + 11 (H_2O) = $C_6H_{22}O_{11}$

The water is used to form the carbohydrates that are in the plant, using calciums as a catalytic base for the plant carbohydrates.

There is a rule that *the higher the carbohydrate content of the plant or the fruit, the higher the mineral content, and the heavier the fruit, the greater the specific gravity.* Any good physicist that knows anything about soil chemistry, if you tell him how much water soluble phosphates are in the soil, he can tell you the carbohydrate content of the produce coming from that soil.

Today, most of the beans produced have less than one percent carbohydrate, when they should have 6 to 7%. Celery has an average of approximately 4% carbohydrate, when it should have nearly 9%. The higher the carbohydrate content the lower the freezing point. The lower the freezing point the higher the mineral content. These are factors in determining the mineral content of foods. The higher the carbohydrate content the more energy it is possible for the person to get from the food they eat. However, in figuring energy we must consider the fact that people have two different kinds of energy. We have the *reserve energy* and the *energy that we use to do our work*.

The two kinds of energy are a rather complicated factor about life.

A newly-born baby girl looks in perfect health, has a reserve energy rating of approximately 8. She would be weighing between six and seven pounds. The reserve energy rating decreases if it is below or above this.

A newly-born, healthy baby boy, weighing between six and seven pounds, has a reserve energy of approximately 10. It generally decreases the same as a girl if weight is above or below. This reserve energy should increase to 100 at 18-21 years and drop off at age 60.

A rather strange phenomenon takes place in both the boy and the girl who use the same amount of calciums until the age of puberty. At the age of puberty, which starts into the age of young womanhood for the girl, she will use seven times more calciums every day than a normal man. If she has the correct amount of calciums she will go into young womanhood with no problems. But if the calciums are deficient she will have all kinds of problems such as menstrual cramps, weakness, irritation, emotional upsets. Anytime a woman whose calciums has dropped too low, starts the beginning of menopause, regardless of the age. It is normal for half the women to be in their late forties or fifties before calciums begin to drop off and the menstrual period stops. There is a place at any age wherein the calciums can drop so low, even in the teens, until they will go through menopause and the natural period will not start again, unless God does a miracle, of which we do not have any instrument to measure the greatness of the love of God. We have to go by the biological, chemical factors which He has given us. Women who have no deficiencies in calciums go through the menopause with no problems, hot flashes, cold flashes, or anything else, and hardly know when their monthly cycle or menstrual period has completed itself.

Therefore when the calciums in women drops, one of the signs is nervousness, easily upset, cries easily. Except for the grace of God, they are hard to live with. If a man doesn't understand this, family life can become one living hell if he has low calciums also. When men's calciums drops low they don't go into menopause, but they are dog-tired at night, or when they come home from work because they have been tired all day. They too have nervous tension. They bark all day and bark at night.

When we are very deficient in calciums it is necessary then to depend on mineral calciums in order to keep our calciums high enough to cause our personalities to be what we want them to be. When nervousness comes about because of low calciums, it causes people to do and say things they wouldn't say otherwise. Low calciums causes extreme nervousness. Nervousness causes people to flare up, scream, take the broom to the cat or dog, because one is fighting for one's life.

Not many doctors seem to be able to help in this situation. However, some very wise doctors do give calcium shots for these conditions. Many medical doctors are afraid to give the shots because they have given them at times with such bad results, and the woman was already nervous and with the shots became a screaming maniac. This was because the wrong calcium was given. Under the present laboratory testing procedures doctors have no way of knowing which of the six kinds of calciums to give. But had the doctor known by way of the RBTI testing, in thirty minutes the woman's nerves would have been calm again.

The type of calcium tests that I have developed does determine which kind of calciums, and under what circumstances, and how much to give according to age, height, weight and sex of the person. I have been conducting RBTI seminars in this for many years. The applicants come from those ministering in the healing arts.

Chapter 6

APPLICATION, WHY AND WHEREFORE

No two people get the same amount of energy out of their foods. Therefore I do not recommend the calorie-counting system whatever. The first reason is that it is not accurate in people that are obese (and less than two percent are obese, and overweight).

I have developed a test called the Reams Biological Theory of Ionization as a result of prayer and fasting for seven days in 1931, that has been used ever since, and has been a great help to people, and today it is spreading all over the world. For this discovery I have been arrested more times than the Apostle Paul. I have not stayed in jail as long, but there are only 44 more states for me to be arrested in for teaching the health message as it is written in the Bible. The Bible says to seek to know the truth, and the truth shall make you free (John 8:32). Also, it says, "Study to show thyself approved unto God, a workman that needeth not to be ashamed, rightly dividing the word of truth" (2 Tim. 2:15).

Today many of the doctors are not permitted by laws made by the drug establishment to do what they know to do to help a person, and therefore, a patient suffers much longer because of these laws of tyranny.

I have been told many times by doctors:

"I have to do certain things although I knew it to be wrong. Everything I do is censored."

"Unless I sell a certain amount of drugs each month I will not be permitted to use the hospitals."

"Unless I send so many patients of mine to the hospital I will be restricted from using the hospital at all."

Whenever there are rules, laws, regulations and union laws that put such restrictions on a medical doctor then we have come to the place that we are paying doctors to keep us sick rather than to get us well.

I have been falsely accused of practicing medicine.

Comparison of RBTI With the Practice of Medicine

There is no logical way to reason, in any shape, form, or fashion, that the Reams Biological Theory of Ionization is the practice of medicine or one of the healing arts. It deals only with food and diet. Everyone is on a diet of some kind even of his own choosing.

The practice of medicine deals with drugs and surgery. Many doctors discredit the idea that food has anything to do with illness. The last I knew, only two of the medical colleges in the United States offer instruction on diet, and these two colleges only give two college hours about diet.

There are no common grounds for comparison between the Reams Biological Theory of Ionization and the practice of medicine, because of the following:

1. The Reams Biological Theory of Ionization is the practice of health through diet.
   The practice of medicine is the practice of illness through drugs and surgery.

2. The RBTI deals with the mathematics and chemistry of a gain or loss of ENERGY, which is determined by accurate laboratory analysis on urine and saliva.
   The practice of medicine deals with diagnosis, which according to Black's LEGAL DICTIONARY, *is a GUESS limited by experience. The guessing is predicated on symptoms.*

3. The RBTI deals with CAUSE and EFFECT due to a mineral deficiency.
   The practice of medicine deals with symptoms and their cure with drugs, surgery, chemotherapy, radiation and cobalt.

4. The RBTI deals with a specific diet for only one individual

and the foods that will bring about an increase of RESERVE ENERGY. This is a natural way to keep anyone's body in the best of health.

The practice of medicine gives no consideration to an increase or decrease in anyone's physical energies, and medical doctors are not trained in any medical college to measure or evaluate energy on a daily or hourly basis.

5. The RBTI never deals with drugs; however, when anyone is ill they should be entitled to the FREEDOM OF CHOICE to choose the RBTI program, or any one or all of the HEALING ARTS to aid them in regaining their health.

    The practice of medicine through the drug establishment seeks to force everyone, even against their will, to use only drugs for their health problems.

6. The RBTI deals with cleansing the temple by drinking clean, cool, pure distilled water, according to or in ratio to weight. The drinking of water has more to do with recovery than the drugs.

    The practice of medicine does not put any emphasis, or seldom mentions drinking water, pure or impure, unless the patient is ill enough to be in the hospital and then they sometimes measure the intake of water and the output of urine. While the patient is in the hospital some doctors insist that the patient drink a prescribed amount of water daily.

7. The RBTI deals with the whole man, the whole woman, the whole boy or girl; such as a tailor-made diet, distilled water, fresh air and supervised exercise according to the reserve energy rating. All this is done on an individual basis predetermined by the body chemistry.

    The practice of medicine deals with diet and exercise for masses of people as though all were exactly alike. Consequently their failures are much greater than they should be because they reject the new as unorthodox.

8. The RBTI maintains that all maladies begin with the organs governed by the vagus nervous system. The cause of the

malfunctioning of the vital organs is that there is not enough mineral in the diet to replace the worn-out cells, causing a malfunctioning or decrease in the production of enzymes to supply other organs with the necessary total daily nutrient to maintain maximum amount of reserve energy.
Those who practice medicine most of the time have no explanation as to the cause of disease.

9. The RBTI maintains that bacterias and fungus are not the cause of disease, however, they will move in when free room and board is furnished. They do aggravate the condition and cause fevers because the body is so deficient in mineral that it (the body) cannot resist their attacks. The RBTI maintains that diseases always strike in the weakest area. The RBTI does advocate the use of vaccines only as a means of last resort.
    The practice of medicine blames bacteria and fungus for the cause of disease. The doctor uses vaccines and antibodies to counteract disease instead of replacing the mineral so nature can restore health naturally. Vaccine may kill the undesirable fungus and bacteria but it does not replace the mineral deficiency, and it also kills the friendly bacteria which protects our bodies from future attack.

10. The RBTI deals with health and happiness and a long, useful life, all well-planned and guided by the RBTI analysis.
    The practice of medicine in most cases deals with suffering, agony, drugs, surgery and a premature death.

11. The RBTI is the adherence to Biblical laws for perfect health.
    The practice of medicine deals with drugs and surgery.

12. The RBTI gives God the glory for the restoration of health.
    Many trained in the use of drugs give man the glory for healing.
    Again this should be evidence that the RBTI isn't one of the healing arts because no one can live without food. I personally believe that the Medical Practice Acts is unconstitution-

al because they are the only statutes on the law books without definition. They do not distinguish between food (diet) and drugs. They do not distinguish between the practice of medicine and the practice of health. They are so broad and vague as to what is the practice of medicine they interfere with one's civil rights in the following ways:
A. The right to exercise the health message as given in the Bible, and the freedom of worship as one sees fit.
B. The right to use scientific discoveries which are patented or copyrighted.
C. The right to make a contract with two or more parties without harassment.
D. The freedom of speech.

Therefore, the Medical Practice Acts, as now written, should be abolished.

The power to license is the power to destroy. A professional license is also unconstitutional because it interferes with the right to work and earn a living. I recommend a registration of qualifications which would entitle anyone to follow the profession of his choice, and the enforcement thereof would be between the employer and employee without the State becoming a scapegoat for a union, church, organization, corporation, etc., etc., which is very costly to the taxpayer.

To locate and zero in on the loss of energy and its causes does not denote the need of a drug. It does denote there is a need of new parts which may be replaced with foods containing those minerals. Therefore the RBTI could not possibly be classified intelligently as the practice of medicine. Since most medical doctors are not trained in human dietary problems as a means of maintaining the maximum amount of reserve energy there could be no intellectual conflict. Greed for money in that they want to stop it would be the only excuse for anyone to try to classify RBTI with the practice of medicine.

Every intelligent medical doctor admits that drugs do not cure or heal any disease. What then does heal?

The answer is the Great Physician, or the Messiah.

How does He do it?

FIRST: by a miracle. Instantaneous healing. He made the laws of health and He can change them anytime He wills for His glory.

SECOND: by obeying the health laws that He has made for our edification. If we do our part the Messiah will do His part, and we shall be made healthy again.

The RBTI is not done by any witchcraft, sorcery, divine revelation, or satanic powers, mind reading, psychic intelligence, or extra sensory perception.

The RBTI test is an accurate means of expressing every degree of biological life in mathematical terms. It is available to those who desire to learn the true science of perfect health through diet.

No one in his right mind would classify diet as a drug. Even children know that food is not medicine. If food is medicine and medicine is a drug then there is no such thing as food. Foods would be only a colloquial name for drugs. Everyone would be a drug addict. All admit that we cannot live without foods, but we can live without drugs.

There is a time for food.

There is a time for drugs.

There is never a time when the abuse of either is excusable. God will not hold him guiltless who abuses the privileges, or consents to be abused by the intemperance of the use of foods or drugs.

No one is down on the use of foods.

No one is down on the intelligent use of drugs.

CONCLUSION: Most people are down on that which they are not up on. Jesus, the Messiah, and the Apostle Paul said these words: "Judge not, that ye be not judged. For with what judgment ye judge, ye shall be judged: and with what measure ye mete, it shall be measured to you again" (Matt. 7:1-2; Romans 2). The medical doctors who call others in the healing arts, or those teaching the RBTI health message, "quacks" should look at their face in a mirror to see if they look like a duck. One thing is sure, if the Bible is true, those who judge others are either a hypocrite or a quacker. Remember, there is no such thing as a seriously dead hypocrite. The dead cannot judge, neither can a dead duck "quack."

I have been practicing health and teaching the health message as it is written in the Bible. This health message begins in the first chapter of Genesis: "I have given you every herb bearing seed, which is upon the face of all the earth, and every tree, in which is the fruit of a tree yielding seed to you it shall be for meat" (1:29). "I will even set my face against the soul that eateth blood . . . For the life of the flesh is in the blood" (Lev. 17: 10-11). In other words, we should take the blood out of the meat. God gave man the right to eat any animal immediately after the flood, but in the eleventh chapter of Leviticus God gave Moses the instruction concerning the clean and the unclean animals. In Leviticus 17 we are given rules for the sanitation of the clean and unclean meats.

The unclean meats will not necessarily keep you out of heaven, but you will probably die and get where you are going to spend eternity a lot quicker.

When I was a young physicist I wondered why God said that a hog was unclean. I reasoned that in that day when Moses wrote that the hog was a scavenger and was around leper colonies, and things of that nature, so was unclean. Today we take a vegetarian and keep him on floor 10 in hospitals, yet we keep the unclean animal and feed him on grain; we wash the fins and he is a "clean" animal. But God says he is unclean. I spent seven years trying to prove that the hog was clean. Do you know what I learned? Absolutely nothing.

I found that the meat of a hog was the lowest of all the meats, not enough to make any such a drastic conclusion, but it is unclean. Nothing could I find that I could say he was very unclean as far as chemistry was concerned. But one day a man about my age (back in the '30's), came into the laboratory and said he had been to the doctor that day and was told that he had cancer and had less than a year to live, and that the cancer was inoperable. He said, "Mr. Reams, medicine has failed and you just have to help me."

This was after I had discovered frequency and had been given the perfect equation. What I did to this man in order to prove both of the discoveries was to give him a gram scale to weigh his foods and told him to take it home and weigh the food in the gram weight. "Eat all you want, whatever you want. Write it down and come to see me every day at 2 p.m. And I mean exactly 2 p.m."

I kept this up for three months, and the man was getting worse. Then I took the three-month data that I had collected and noticed that every day that he had eaten ham, bacon, sausage, pork chops, spare ribs, shrimp, lobster, tuna, mackerel, oysters, or foods of this nature, that the energy rating decreased faster than it did the other days. I took him off all these unclean foods, plus iced tea and chocolate, and put him on a diet. He immediately began to recover. He is still living. I talked to him in July, 1977.

From this I learned, after studying another two or three years, that the unclean meats digest on an average in about three hours, which releases too much heat energy in the body and causes one to grow old too fast. Then I began to advocate that one not eat the unclean foods whatsoever.

I have seen while doing work for medical doctors in hospitals, patients who looked 70 and 80 years old and were in their late thirties or early forties. We would take them off of pork and unclean meats, and they would get well in a hurry. Also, in six months they looked even younger than their years. It was amazing just to see that God knew what He was talking about 4,000 years ago when He made these rules.

After my own children were born I began testing them. Not that I wanted to harm them or try things on them, but to know how to make it better for them to really keep them healthy. I wasn't experimenting with them with anything that could endanger them. It was strictly with foods and minerals. When I discovered the energy level of babies, I did so with my own children.

I also discovered that babies' gastric juices are so diluted, so weak they cannot digest the foods that adults eat. By mathematical calculations I worked out the foods that a baby could digest from the very earliest time that it could take food until its gastric juices became strong enough to digest the foods of an adult. Consequently, my children did not get any nut, or nut meats, or nut butters (coconut being the exception) until they were eight years old. (Or, unless the nuts are steamed soft in a pressure cooker, or boiled until soft. If more parents would do this they would have healthier children.)

I also discovered children could not digest meats until they were 12 years old, therefore ours didn't get any meat until they were that age. We raised six children with no cavities in their teeth, and no days missed from school because of illnesses.

Many times people visiting our home would say: "How can you eat meat before your children without giving it to them?" The children never questioned it because they were told that meat was for big people and vegetables and fruits, soya meats and soya products were for children. The children would have the soya turkey, soya chicken, linketts—all made from soybean. They had the Numeat and the Nutmeat from many soya products to get their proper balance in diet. Nearly every morning was a fruit breakfast, except once or twice a week they had an egg for breakfast, either with millet cooked soft, or grits that was cooked soft, or oatmeal. Every morning of the week they had a different breakfast. Even in a month they had a different breakfast every day. They had a great, great variety of foods. They were happy children. They were never fussy. When they became a little cross I would run a test on them and find out why—if it was due to a mineral or vitamin deficiency. We would correct that before we punished a child—which was seldom. It was very rare that our children became cross, or hard to live with.

Besides meats and nuts the children should not eat shell fish, oysters, clams, lobster, or any soups with meat or meat broth in them. Children cannot digest chocolate, iced tea or coffee. Our children never tasted coke, or carbonated drinks. We always had fresh fruit drink. There is a time and a place under medical dietary conditions that there would be a place for carbonated drinks, but it would be very, very rare.

When we visited homes where they had meat on the table our children would just say, "No, thank you. I don't care for any meat." They just ate the vegetables.

When children came to our home we had a table for the children and a table for the adults. On the children's table they might have a soybean roast, soybean burgers, or steakettes—all made from soybeans. Our children did not know that they were not eating meats. They were not so concerned with what they ate, it was whether it tasted good or not.

We taught our children very early in life, by the time they could chew anything, to eat leafy salads. We also made green drink with a juicer. We ground our own grains and wheat and made two loaves of fresh bread each week. That is all the bread we had for a family of eight. We used perhaps six pounds of potatoes *a year,* because potatoes are one of the poorest of all foods, lowest in mineral, lowest in everything. If it were not for the gravy, or the butter, or sour cream, or things you put on a potato, we wouldn't even eat it. We did use potato buds in our home occasionally for thickening of gravy, it was used very sparingly. We most often used egg plant to thicken the gravy. We put egg plant in small containers in the freezer, and when we needed to thicken the soup we would use the egg plant.

We ate our high carbohydrates for breakfast. We had the biggest meal in the middle of the day. Even when the children were in school they took their big meal in the middle of the day. We gave them breadless lunches. Only one day in the week were they given bread in their lunches. At night we would have soups, and maybe a slice of tomato, a piece of toast, a cup of herb tea, or yogurt. Mostly plain, homemade yogurt.

Chapter 7

CAUSE OF DISEASES

The first day that anyone, after the age of 18 to 20 fails to take in as much *reserve energy* as they burn up in *regular energy*, this is the first day of illness. This first day of illness is caused by mineral deficiency.

Parents should pay very close attention to children's complaints. One little boy was sent to me by a medical doctor, crying because of a stomachache. The boy was in the first grade in school. He had cried and cried for months about his stomachache. The doctors couldn't find a thing wrong with his stomach. When the child was brought to me he was in the zone for a fatal heart attack. It was his heart hurting him and not his stomach. He was having chest pains and he thought it was his stomach. When given a diet to correct this condition there was no more problem.

One of the sad things about America is that over 16,000 children drop dead on the school grounds every year because the proteins and the foods they are eating are too rich for their bodies. It builds up to the point that it puts the body chemistry in a zone for a heart attack.

A heart attack just doesn't jump upon anyone. An electrocardiogram (EKG) does not tell you if you are going to have a heart attack. It only tells if you have had one. A pectoris heart attack is caused by the diet being too rich for the child, and a child cannot digest the proteins. The undigested proteins turn to urea and it causes the child's cells to be exchanged too rapidly. The child is not drinking a sufficient amount of water to wash out the excess urea solids and it causes the heart to beat too hard. In a stethoscope it sounds like a drum, it is beating so hard.

Because the heart beats so hard so long until it gets so tired it takes a rest. More children die with cardiac arrest than a heart attack. It is very difficult for the doctors to find out why the child died. We don't know how many children get to a hospital and die, and how many get home and die later because of cardiac arrest. All of this can be prevented by the RBTI tests that I have developed.

Ninety-five percent of the crib deaths could be prevented because many times the proteins in the mother's breast milk are too high for the child and his little heart beats too hard and the heart just stops, and the mother finds the baby dead in the crib. I have tested many babies under the supervision of doctors and found that they were all in the zone for a fatal heart attack. This can all be prevented. But it has to be handled through the dietary process of ionization. At the present the doctors do not have any of these tests available to them, unless they have taken the seminars that we teach for those in the healing arts.

On adults tests have been made that indicated that they were in the zone for a fatal heart attack. We had one case in which we told the man he was in the zone for a fatal heart attack and asked him to drink four ounces of water every half hour for eight hours while lying down and doing absolutely nothing and then return for another test. He made all manner of fun of us and said he wasn't even sick. This test was made between nine and ten o'clock in the morning by a medical doctor. At five o'clock that afternoon his wife called saying he was pronounced dead on arrival at the hospital.

We have made tests and predicted heart attacks anywhere from 30 days to six months and missed it by just a matter of hours. We predicted one three years away and missed it by less than three weeks. That was when the man would do nothing about it. Energy can be measured and we can know when it is going to drop below what we call the PNR line—the point of no return—and once it drops below that line it is just a matter of hours, or even minutes, until death will take over.

There is only one cause of death, and that is old age. It doesn't matter if a baby is born dead, or if it lives to be 150. Old age is nothing more than that the body does not have enough energy to maintain life. In mathematical terms this is all death is.

You can bring it on early, or you can delay it.

There are also some factors that are very important in diet. At age 20 it takes x amount of reserve energy divided by time to make a cell in the body if there is reserve energy behind it. At age 40, if the reserve energy is still 100 it takes two times the amount of reserve energy divided by time to make the same cell. At age 60 if it is still at 100 it takes three times the amount of reserve energy divided by time to make the same cell. At age 80 it takes four times the amount of reserve energy divided by time to make the same cell. So aging is how long it takes nature to make a cell, and how much reserve energy it takes.

I find people who are 60 years old and only use 1½ amount of reserve energy, or 2, or maybe only three times the amount of reserve energy. I found one 80-year-old who used two times the amount of reserve energy. He was like a 40-year-old man.

The process of aging can be delayed by an intelligent diet, proper exercise, and proper thinking. Diet has more to do with thoughts than thoughts do with dieting because the weaker one is, the more difficult it is to think intelligently, the more incapable one is of deciding what is good and what is bad for them. The weaker one becomes the less thoughts they have about facts. This is in agreement with the Bible because in it we read: "Faith without works is dead." People who are sick have less faith than those who are well, because they can't work. The healthier you are the greater the possibility for faith or self-confidence, and the greater your works. It is very important to relate the Scriptures throughout the entire Bible to diet.

Can you imagine Jesus getting up in the morning with a bad cold and sneezing all over heaven, or have the lumbago, or arthritis in the joints? We are created in His image. He is in perfect health and He wants us to be in perfect health also. And we must be if we are to be like Him someday, as Adam was in the Garden of Eden before the fall.

It is very easy to stay healthy. But if you get sick you really have to work at it. You have to break all the rules. You have to use your stomach for a garbage can. You have to work like a machine, never stop and never rest. You have to have hate in your heart. You have to break the Golden Rule. You have to really make a jackass out of yourself.

Three years ago I bought a very good set of medical books. I paid $800 for my daughter to use them in helping me teach the seminars in the RBTI. This would help her to know something about the names of diseases and problems of people. After reading the first book she said, "Daddy, all it tells you in here is how to name the disease. They don't tell you what to do for it."

"Honey, I know that," I said, "then you have to look in the PDR (Physician's Desk Reference) in order to find out what to do for the disease."

Today the drug manufacturers tell the medical doctors what to do for the disease. All the doctor has to do is to name it, and it is his hide if he doesn't do what the PDR says to do.

Naming the disease doesn't cure anything. Diseases are caused by mineral deficiencies, and no drug will cure anything. For example, the swine flu epidemic was predicted three years ago by the drug manufacturers. Consequently, the swine flu vaccine was completed, never tried or tested by HEW, or the Food and Drug Administration (FDA), and it was given to many, many people. It killed some, crippled others. Yet laetrile has been invented now for 15 years or more, and it is being condemned by the FDA because they didn't think of it, and anything they don't think of isn't any good. They never tested it, neither has the American Cancer Society done anything about it. I have had two definite cases wherein laetrile was recommended by medical doctors and given under their supervision, wherein I had the privilege of doing the tests daily on the persons. I discovered that the hydrogen cyanide in colloidal form, which is not poisonous, made vitamin C available in a very high amount, and the patients both got well. One was a cancer patient and the other a muscular dystrophy patient.

Many doctors are so afraid of the drug establishment until they are not at liberty to advise patients, or people who are ill, for fear of having their license removed. The license of the medical doctor is illegal, it is unconstitutional. Licensing a person does not make him any smarter or give him any added ingenuity whatsoever.

Any profession that needs law to keep it in existence is opening the door for that organization to be corrupted. This is true of any profession that is licensed because if the profession was operating 100 percent legitimate it would not need law to protect it. If it were good the people would flock to their door, and if it were bad, the practitioner, either legal or medical, or any other profession that was kept in existence by law, would starve to death.

Once any profession becomes legalized then it means that it has to have law enforcement agencies and these agencies become dictators of the establishment, which has made many of our courts the courts of injustice. Many of our hospitals are now the shortest route to the cemetery.

When a doctor says that a disease is incurable, he means that he knows of no drug that will cure that disease. He doesn't mean that God can't do it. So what we are teaching in our RBTI seminars is God's way of getting people well. In II Chronicles 16: 12-13 we read: "And Asa in the thirty and ninth year of his reign was diseased in his feet, until his disease was exceeding great." This was probably diabetes. The verse following this says: "Yet in his disease he sought not the Lord, but the physicians. And Asa slept with his fathers, and died in the one and fortieth year of his reign." God would heal many if we would just obey His rules, instead of depending on drugs.

In the Book of Revelation we are given some very prophetic words: "Neither repented they of their murders, nor of their sorceries, nor of their fornication, nor of their thefts" (9:21). In the Book of Revelation the word *sorceries* means drugs, drug salesmen, or drug purchaser, or drug user. It is the only place in the entire Bible that it means drugs. It comes from the Greek word *pharmakeia,* meaning drugs.

In Revelation 18:23 we read: "And the light of a candle shall shine no more at all in thee; and the voice of the bridegroom and of the bride shall be heard no more at all in thee: for thy merchants were the great men of the earth; for by thy sorceries were all nations deceived." For by their "drugs" all nations were deceived. Also, in Revelation 21: "But the fearful, and unbelieving, and the abominable, and murderers, and whoremongers, and SORCERERS, and idolaters, and all liars, shall have their part in the lake which burneth with fire and brimstone: which is the second death." Here we are told what is going to happen to people who use or sell drugs. In Revelation 22:15 it says: "For without are dogs, and SORCERERS and whoremongers, and murderers, and idolaters, and whosoever loveth and maketh a lie." John, who was on the Isle of Patmos prophesied about the deceiver of the world by drugs. We are now living in the midst of this generation of drugs and drug abusers. There are even stores that are called "drug" stores.

I'm not speaking against the very fine Christian physicians. They are wonderful men of God, but many are so coerced and legalized until they cannot do what they know to be best. This is the advice I give to my readers: Any doctor who cannot give you an improvement in three days—then it is time to get a new doctor.

Any doctor that claims that cobalt, chemotherapy, or radiation heals is a deceiver. Please get away from him as fast as you can. None of these things have ever healed anyone.

Don't keep going to a doctor and get worse and worse because you can do that at home for nothing.

On January 23, 1977, I was poisoned with laudanum. I was in the hospital three days and nights and did not know at that time the doctors had tested me with practically every instrument they had in the hospital. I was going all hours of the night for tests. I went in on Sunday afternoon at 5 o'clock and on Wednesday morning two doctors came in and said: "We can't find anything wrong with you."

"I'm dying and I know it," I said. I was waving in and out of consciousness, rather rapidly. When I did come back my mind was as clear as a bell. I found out later that they had run $4,434 worth of tests and couldn't find anything wrong with me. I then said, "I don't need you any longer, if you can't find anything wrong with me when I know I am dying."

My former pastor brought in a Korean doctor. I gave him a rather hard time in and out of consciousness. He said, "Give me 24 hours to find out what is wrong with you."

The other two doctors refused to let my daughter come in and run a RBTI test. But Dr.–, to satisfy me, permitted my daughter to come in and run the test. She was working and couldn't come in until seven o'clock that night. She ran the test. The next morning when the new doctor came in he said, "I can't find anything wrong with you."

"Doctor," I said, "I have found something wrong," and told him where it was. He looked there and found it. In two hours I was on the operating table. They took over a pint and a half of pus out of two blisters over the colon. My white blood count (WBC) had risen from 7,400 on Sunday to 19,000 on Thursday morning, indicating that I was in serious shape. I got better then.

This new doctor wanted to take the RBTI seminars. It was available in California. "If you go to California, can I go with you?" he asked.

So we went to California by plane on a Sunday for I was still very, very weak. On Friday morning I had another bad withdrawal from the laudanum and had to go into the hospital in California for 12 days. Then I went back to Orlando, where I had to go through more surgery in order to correct the damage that was done to this point. It burned so seriously, and yet laudanum is pure opium and alcohol. The mistake made in poisoning me was that they mixed it with wood alcohol. My body rejected the wood alcohol, and I vomited up most of it. Which was such a rare thing for me.

I did exactly what I am recommending you readers to do. If a doctor can't show you any improvement in three days, get another doctor. Let surgery be the last resort.

I do not believe in vaccinating children at all. Give them the proper amount of water to drink (divide their weight by two and call it ounces, and that is how much water a child should have to drink each day, and it should be distilled water), and there will be less illness. If everything else fails use vaccinations as a last resort. If you are in doubt about what I am saying, there is a book in print, *The Poisoned Needle,* read it.

I never permitted my children to be vaccinated. When it came time for them to go to school I was told that they had to be vaccinated. I told the principal, "My child will not be vaccinated, because it is your job to educate and not to medicate. It is a federal law that my child must go to school. And unless that child goes to school I will enter suit against you and the school board for practicing medicine without a license."

"I will call you about it," he replied.

In two days he called and said, "Send your children on to school, you'll never have any problem with us about vaccination."

Today America boasts on having the finest hospitals and the finest doctors in the world and yet we are 37th in the list of nations having poor health. What a shame to crow about our excellent hospitals, our fine doctors and having the nation 37th in the list of poor health. It is like a fellow in his late teens I knew that was a moron. He was having trouble finding a girl friend. One day he found one and had his first love affair and contacted gonorrhea. He went out boasting to the fellows, "I got it, I got it." He had the proof of it—he thought he was supposed to have it. Now for us to boast of our hospitals and fine doctors is boasting of our illness. Shame on America.

My goal is to make America the healthiest nation on earth, and for this reason I am falsely persecuted for telling the truth. Evidently the truth is not allowed in our courts any more.

We are paying our doctors to keep us sick rather than to keep us well. You might ask why this exists. The reason is that there is no money in good health. America, where is your sense of values? Do you take pride in poor health, or do you take pride in good health? If enough of our people become ill our nation will become endangered because sick soldiers cannot win battles. Neither can sick people resist temptation, and the family is weakened and the nation is destroyed, all because of poor diet—which can be corrected by the Reams Biological Theory of Ionization.

Chapter 8

CAUSE AND EFFECT

Biological things, or everything that has life that moves or grows is because of the harnessing of energy. Energy has to be harnessed in order to accomplish the desired results.

We have established a cause and effect way of measuring energy in human beings. That is how the reserve energy is determined, by resistance, as all other energy is calculated.

Without resistance energy cannot be created. Resistance is between anions and anions, cations and cations, or anions and cations, or we can say, between acids and alkalies, or negative and positive electricity. It is the resistance that makes it possible for a potential synchronization, or a levelling point which will establish the frequency in a human being. God created us with this frequency. We cannot do anything whatever to change this frequency. Even when a body is burned to an ash, the ash will still bear that frequency.

We are dealing in this chapter with adult, healthy people only and not sick people. We are dealing with rules that if disobeyed will lead to illness. Anyone can be well, but in order to be sick you have to work at it—here then is the basic reason for cause and effect.

No one can break the biological laws of physics, or disobey the laws of God without bringing themselves to grief, illness and an early grave. One of the things that never ceases to astonish me is the great number of people that are in such a hurry to get to the cemetery. They just can't wait. They break every known biological rule of health, of physics, and of theology.

Even if you are in perfect health, and have a reserve energy rating of 100, and you break the Golden Rule, you then have made the first step toward illness. DO UNTO OTHERS AS YOU WOULD HAVE THEM DO UNTO YOU. Healthy people are happy people. You are not healthy if you are continually seeking revenge, or breaking the Golden Rule. If you break the Golden Rule you first must have to hate yourself. *Hate is the finest cancer seed in the world.* Hate leads to greed, and greed leads to revenge, and all of these things have an effect on the digestive system. The food doesn't digest as it should. Consequently, your rest is disturbed at night. The next thing you notice is that you are more irritable than is normal and you wonder what is happening.

The first thing that hate does is to prevent the liver from taking in oxygen to assimilate the calciums and other minerals in our food, and the lack of calciums is the beginning of a diluted gastric juice, which decreases our energy.

In dealing with diets we have to deal with the whole man, the whole woman, the whole boy, the whole girl.

Another sign of ill health is when people lose their desire to sing. Healthy people sing. Therefore they breathe deeply from the bottom of their lungs. The Bible tells us to sing, or at least to make a joyful noise unto the Lord. It didn't say carry a tune, so the tune is not important. But singing is a sign of happiness. And happiness is a sign of good health.

There are many facets and telltale things that denote that we are losing energy, or that our reserve energy is not as high as it should be.

The lower our reserve energy, the more ill we become. Now this is cause and effect.

Mineral deficiency is one cause of loss of energy. Mineral deficiency can result because of hate, revenge, the breaking of the Golden Rule, the lack of proper digestion. Ill health does not result from one single event in our daily life, but a combination of things, plus lack of minerals in our dietary program.

When we fail to think positive thoughts, wallow in the mire of gossip, refuse to forgive and forget, then our bodies are becoming a mental garbage can and it cannot be filled with the Holy Spirit. Things that do not exalt, things that do not lift up, but tend to tear us down should be avoided. If we are to be healthy we must first follow not one example of living, but many.

I have seen people that were very ill put on the diet and for two or three weeks their bodies did not respond, yet there was no genetic reason in the test made on the urine and saliva as to why their bodies should not respond. I would go to these people and ask, "Tell me about your childhood. Tell me about the people that you loved most, tell me about the people that you hated." After some thought they would begin to tell me about the people they liked most and the people they hated. And they STILL hated those people, even after 50, 60, 70 years. Hate, even though it is almost forgotten, still affects the digestive system.

We must choose the foods recommended by our counsellors, doctors, and chiropractors who have studied the RBTI in order to increase our energy.

If you would look at the atomic chart of the specific gravity of the elements you will find that hydrogen has the specific gravity of one; (I am giving these figures in whole numbers) but helium has a specific gravity of four; lithium has a specific gravity of six; nitrogen 14; oxygen 16; phosphorus 30; potassium 39; calcium 40; manganese 54; iron 55, etc. All of these elements and many more are necessary to maintain life.

The higher the specific gravity of an element the stronger the gastric juices must be (the bile from the liver), in order to extract and assimilate from our diet the heavier elements, or elementary energy, from the food. The higher the specific gravity the greater the weight of molecule, or the greater the weight of any given amount. For example: one cubic inch of iron would weigh a little bit more than one cubic inch of manganese. Iron would weigh much more than calcium. Calcium per one cubic inch would weigh more than potassium, and much more than phosphate. Oxygen would be heavier than nitrogen. Nitrogen is heavier than helium. Remember, we do not live on the foods we eat, we live off the energy from the foods we eat.

The energy is created in various organs at a varying ratio of different elements. For instance, the **brain** contains more potassium, that is, it is higher in the brain cells. The **liver** contains more calciums than anything else. The liver contains the greatest amount of calciums but the next highest is iron and iodine. The **heart** contains much calciums but the next very essential element is arsenic. (If there is enough arsenic in colloidal phosphate form in the diet you will not have heart valve trouble, heart weaknesses or heart skips, which develop in later life.) The total **sex organs**, male and female, contain more manganese than any other elements. The manganese is the one element that contains the power of life. Without manganese there would be no life upon the earth. It is the key factor of all the elements which is responsible for life to continue. An excellent example of manganese deficiency can be seen when you open a peach seed and the seed spreads open. But if it is a severe manganese deficiency there will not be any seed inside of the stone or pit. It will be decayed, or just a bit of pulp with no life in it. Another example is a peanut which has a shell but has nothing in it (called pops), or a pecan with the meat of it greatly dehydrated, or half of it is black, or has no meat in it at all.

Whenever people do not get enough manganese from the food they eat then it is the first sign of the beginning of a loss of energy, and delta cells and omega cells (carcinoma, or cancer) are beginning to occur in these organs. In the medical books there are no terms for perfect cells, only terms for carcinoma, or cancer cells. We have made some new terms of names for cells in order to say more accurately what we mean. An ALPHA cell is a cell with 100% of its reserve energy. A DELTA cell is an in-between cell that has lost some of its reserve energy, it is wearing out, but has not yet worn-out. An OMEGA cell is a cell with no reserve energy, it is worn-out like a tire that is absolutely impossible to use any more, cannot be patched up, is beyond use. In this way we can zero in on the exact loss of energy because the delta cells are losing energy but the omega cells are not losing reserve energy, they are already dead.

A deficiency in potassium over a period of time could lead to a brain tumor.

A deficiency in iodine and iron in the liver, as well as calciums, and deficiency of oxygen in the lungs could eventually lead to cancer of the liver, or any other organ, if this condition is permitted to continue over a long period of time. The weakest part of the body will be affected.

If there is a phosphate or calciums deficiency it could eventually lead to cancer of the bone, and also cavities in the teeth.

There can be two reasons for the mineral deficiency. One is that the gastric juices are too dilute to digest the chlorine, fluoride, calciums, and phosphorus which are necessary to make the enamel in the teeth. The second is even though the gastric juices are strong enough to assimilate them, the foods may not contain these elements. In the RBTI the tests denote these deficiencies and are shown in two ways:
LOSS OF ENERGY.
WHERE THE LOSS OF ENERGY IS.
Consequently, the RBTI tests can predict as much as ten years in advance where cancer will strike, or whether or not the person is going to have a heart attack, or whether or not there are going to be health problems. The deficiencies may linger for a long time. This is explained further in the discussion of the numbers which denote these deficiencies.

One of the great problems today is the deficiency of minerals in the foods that we eat, whether it is fresh, canned, frozen or whatnot, it is a deficiency, or has a lack of mineral. I am going to lay this blame strictly where it should be: the agricultural experimental stations. I'm not casting reflections on any one person, all I'm doing is stating a fact for most of our textbooks on agriculture are written by the experimental stations under the supervision of the fertilizer and the insecticide companies.

Most of our college graduates in agriculture today learn to do three things: to buy fertilizers, machinery and sprays.

I was in the agriculture engineering business for 38 years, and had 32 engineers working for me when I retired in 1968. We were teaching farmers how to get a greater yield at less cost per acre. But the experimental stations fought us in every way possible. They never missed an opportunity to slander the work we were doing. However, our work prospered, and we were never able to catch up with all the requests we had for help.

The average farmer in the United States today is producing less than 25% of what that farm is capable of producing. The greater the yield per acre the higher the mineral content. The higher the carbohydrate content the higher the yield per acre. The higher the yield of an acre the less time it takes to produce it. **High quality produce with a high carbohydrate content will not rot, it will dehydrate but it will not decay.** A few years ago I showed in a fair in Orlando, Florida, for three consecutive years the same watermelons. They were marked and kept under supervision and under the eye of the county agent. The melons had been picked eight months before they appeared in the fair the first time.

The most successful farmers today have their own business, and their own soil scientists to guide and direct them, and they are not giving out this information to anyone whatever. A very simple example is this: the difference in hard wheat and soft wheat is solely in the mineral content that is in the soil. If the mineral content is high enough in the soil it is hard wheat. And if it is not high enough it is soft wheat. I know there are wheat growers who will disagree with this. However, 100 years ago much of that same soil produced hard wheat, and today it is producing soft wheat. This fact cannot be denied.

Many corn growers used to produce 200 bushels of corn per acre, and some of that same soil is down to 50 bushels yield per acre now. This is because the mineral has been taken from the soil and has not been replaced. The Bible prophecy is true: the earth shall wax old like a garment. Men shall become weaker but wiser. These minerals are available, and are very economical to put back into the soil to help make American people more healthy.

Any farmer who is not producing 200 bushels of corn per acre, two tons of peanuts per acre, 20 tons alfalfa per acre, or 20 tons of watermelon per acre should be ashamed.

We are teaching seminars in agriculture to teach the farmer how to produce a greater yield per acre at less cost per pound or per capita. This system works, and is proven.

No one person, no one doctor, no one dietician, no one merchant, no one farmer, no one dairy farmer, no one meat producer is responsible. It is a chain reaction that is cause and effect of our entire society regarding good health. Therefore, if we want to be healthy the best way is to use a very, very great variety of foods.

In the laboratory tests that we do, the Reams tests are standard laboratory tests, made by standard laboratory instruments. I was the first man to apply these biological needs to the functioning of these instruments. The Lord taught me how to do it, and this is the forty-sixth year that this system has been in use, and has been proven accurate.

One of the problems that we have been facing for the past 21 years is the state laws made by the drug establishment, which has greatly restricted this knowledge, or greatly kept this knowledge from the people. Consequently, I have been in hot water so to speak with the drug establishment. I have been fighting them in the courts now for nine years. And the battle still continues.

One of the great things that is going for the American people is the National Health Federation, the International Cancer Society, and the International Association of Cancer Victims & Friends. These have done much to help in the fight for the freedom of choice and to enlighten the people on their rights to choose whom to go to with their health problems.

Humanity changes very slowly. An example of this is Moses leading the children of Israel from Egypt into the promised land when they were marching through the valley of snakes and serpents. Moses put up on a pole a serpent and said to the people that all who would look toward the serpent on the pole would be healed. Most of the people refused to look because they said it wouldn't do any good. So humanity today hasn't changed very much. They know to do better than they do. They are told what to do and they won't do it.

Many people come with dietary problems. They get tested and the moment they think that they are going to live they go back to the old system until they again think they are going to die. Then they get tested again and get back to the point where they think they are going to live, and over and over they go through the same process. Never well, never sick enough to die. This is human nature.

It is quite disheartening to the doctors and the testers who know that they are able to help get the people practically well and the people then not seeming to want to be perfectly well. They just want to be able to think they are well. They want to do what they want to do.

It is an interesting fact when Moses was leading the children of Israel out of Egypt, in two years they arrived at the promised land. But the oldtimers were afraid of the enemy. They were afraid of change. They didn't want to rock the boat. Consequently they didn't get into the promised land. So for 38 years God let them march in the wilderness, until all the oldtimers died off. Then God took the young people into the promised land. Even so today, the RBTI probably will be accepted by the young people.

People who break the rules of health grow old prematurely. So naturally they go to a premature grave. There are many names for the diseases that caused them to die. However, the cause of the death was a mineral deficiency.

All illness is a result of a mineral deficiency. No drug can supply a mineral deficiency. The reason most people get well in the hospital is because they rest. Some illness comes about because people do not rest, and lack of rest depletes minerals. The only way health can be restored is to supply the mineral needed for God to restore the decayed tissues.

If we will supply the minerals, God will do the rest.

A famous doctor once said, "All I do is entertain a patient while nature heals."

In the process of healing there is a divine process and a miracle in itself. We cannot do miracles. We can only set the stage for God to do His miracles.

One of the saddest things about America is the fact of how much we are abusing our children. The public schools are often giving them foods that their bodies cannot digest, and they are getting diets all alike regardless of whether or not their body chemistry is synchronized to accept it. If the body is not synchronized to take food then the food moves the body chemistry out of the range of perfect, and the further from perfect range the body chemistry gets, the more ill one becomes.

It is not practical for every child in a public school to have the exact same food. Some have diabetic tendencies, some have low blood sugar tendencies, and some have digestive problems, some are overweight, some are underweight, some have worms, some have hypertension problems, and all of these can be corrected by diet. But in order to do that, tests have to be made. There are three groups of food that will fit all categories of children's diets that can be served in any school.

My dream is to see that the RBTI be used in all schools, and diets made to fit the children by groups instead of each child choosing what food he wants in his lunch. And also helping parents in the home to supply the need of the children.

The purpose of the RBTI test is to determine whether or not the body chemistry is too acid or too alkaline, whether it has or is accepting too much carbohydrates, or whether it is not getting enough, whether the body is retaining too much salt or not enough, whether the body is digesting the proteins or not digesting the proteins, etc. It is God's plan in bringing anyone's health back to perfect. The RBTI works well on all people whose body chemistry will receive or will respond to diet. There is less than 1 percent of all the people who are placed on a diet to bring their body chemistry back into a perfect range whose body will not respond. This is because there is either brain damage, a tumor of the brain, a potassium deficiency, or low blood sugar in which the brain is not getting enough oxygen, or brain injury or damage to the central nervous system (which is the nerve that takes the message from the brain to the vital organs).

This plan of diet is not the practice of medicine in any shape or form, because everyone is on some kind of a diet, either good or bad.

Some people are inhibited by cigarettes and some are not.

Some people are inhibited by drugs, 80 percent given by medical doctors. (The other 20 percent of the people on drugs buy them on the black market.)

Some people are inhibited by diet. Their body accepts too much of one thing and not enough of the mineral and vitamin content of another.

All the RBTI diet does is correct a person's mineral and vitamin deficiency. If the person gets well, it's not my fault or anyone else's I have taught. God does it. *No one can heal another; only God can heal.*

There have been some very simple instruments made that will be on the market soon that the housewife may use. Instruments that will be easy to use and read, as a watch dial which will show the mother how to prepare a diet for her family. If there is some member in the family that needs a different diet, these instruments will show her how to make the diet for this particular individual. This will be one of the greatest factors in the world to prevent family illnesses. This will give a family more money for pleasure and the things they want to do, and to be healthy. It will decrease the loss of man hours due to illness by hundreds of percent.

More progress has been made in every field of science than has been made in the field of health for the people of this world.

The purpose of this book is to enlighten the people that it is easier to be healthy than it is to be sick.

Chapter 9

REAMS BIOLOGICAL THEORY OF IONIZATION

The theory of ionization was discovered after a father with a three and one-half-year-old child had been to a medical doctor. The child was having seizures and was diagnosed as an epileptic. The doctor had told the parents that this child would go into a seizure and would not come out and could not live to be five years old.

"You just have to do something for my child for drugs have failed," the father of the child said. They lived two doors from us. I knew the boy well.

When I went to the laboratory that night I just sat for three days and nights wondering where in the world to start. What could I do? What approach should I use? Mother would bring my meals to the lab, and from one meal to the next I would not have touched the previous meal. The days almost seemed like minutes as I was meditating around the clock. I didn't sleep at all, I was so engrossed in the problem at hand. I was not even sleepy.

On the end of the third day the thought came to me that if I knew the mathematical chemistry of perfect health, then tests could be run on the boy's body chemistry and we could find out how far from perfect his body chemistry was, and then we could make a diet and bring his body chemistry back to perfect.

It was then I started from the knowledge of frequencies I had on grapes. It was the only thing I knew the frequency of at that time to calculate, or you might call it "dead reckoning," what a human anatomy should read if it were perfect. I'm sure that an angel held my hand for in four days I had come up with a formula which I consider today to be perfect for human anatomy, regardless of the age of the person. This was almost four years before I had discovered the frequency of the human being, which proved that this equation was indeed correct.

I asked the parents to bring the little fellow in and we'd run tests on him. The first equation was much longer than the equation we have now. We ran tests on stool, blood, tears, ear wax, finger and toe nails, sweat, hair, urine and saliva. As far as I know, I was the first person to do a hair analysis. After a week I had completed the test and was amazed at how much duplication there was. I began to drop off various things in the next three years until I had left only the equation I now use.

We had the parents come in and made a diet for the boy, and found he was not an epileptic at all, but that he had low blood sugar, which was causing the seizures. His pancreas was manufacturing too much insulin. We corrected the diet by using fresh green, raw chlorophyll. We used some St. Augustine grasses and ground them in a food chopper, put them in a strainer and pressed out the juice with a fork. We didn't have juicers at that time, as they had not yet been invented.

Immediately, the boy began to improve. In three months there were no more seizures. In a year he was doing well. Then the family moved to Texas, and I didn't see them again for 35 years. I was on the streets of Orlando, Florida, near the corner of Church Street and Orange Avenue when a young man came to me and said, "Are you Carey Reams?"

"Yes," I said.

"I know you, but you do not know me," he said.

I said, "No, sir, if I've ever seen you I don't remember."

Then he told me, "I'm the little fellow who had the epileptic seizures, and you gave me a diet. I remember you real well. I never had another seizure. When I was seven years old, we were in a little accident and I had a light seizure and went into shock. But I've never had another since that time."

This was my first test of this ionization theory.

From time to time I was called upon to do this test. I had talked to different doctors about it. At this time I still had the medical laboratory, and doctors began to use it some and found the test to be accurate.

When I discovered frequency it was brought about by two police officers bringing some ashes from a building that had been burned. They had reason to believe that the ashes were from humans, or possibly an animal. The undertaker had a crematory (he and my father had been friends). I had known him all my life. I went to him and told him my problem, and asked to borrow some ashes of a corpse where he knew the sex, age, height, weight, and color of hair. In two weeks I had established the frequency upon which human beings live, and could even tell the race, but could not tell anything about the age because the age of carbon is measured in thousand of years. I returned to the crematory every atom of the ashes.

I was able to determine that there were two females and one male in those ashes. I did the work for the police department for a number of years, until I went to war and then I gave it to the Commission. They don't follow it any more because it is an aggravation to sit in court for a few days waiting to testify for the little bit you get for the testimony of it. Police departments today can determine if it is human or animal ashes. This was one of the steps in which the RBTI was discovered.

I sold the medical laboratory, and then spent my time in the Agricultural Engineering field. I knew I had something, but did not know where its strong and weak points were. When any chemist or physicist makes what he thinks is a discovery, the first things he wants to know are:

Is it real?
Is it true?
Does it work every time?
What are the exceptions?

I did then what is the duty of every scientist and physicist: set out to prove that this theory was not true.

The theory of the entire equation is:

Whether the body has too much carbohydrate, or not enough.

Whether the body chemistry is too acid, or too alkaline.

Whether the body is retaining too much salts or not enough. (There are many different kinds of salts. Chloride salts, and salts without chloride. Even insulin is a salt.)

Whether the body is throwing out the worn-out cells, or not throwing out the worn-out cells.

Whether or not the proteins are digesting.
Whether there is too much manganese, or not enough.
Whether there is too much iron, or not enough.
Whether there is too much iodine, or not enough.
Whether there is too much arsenic, or not enough.
Whether or not there is enough potassium, phosphates, etc., of all the elements.

The perfect equation is:

1.5 (1) $\underline{6.40}$ (2)   6-7C (4)   .04M (5) $\underline{3}$ (6)
            6.40 (3)   3 (7)

The first number denotes the carbohydrates (1.5).

The numbers (2) and (3) denote the lack of calciums, or the resistance between the anions and cations, the anions and anions, cations and cations. This is a key factor in measuring the total amount of energy in our bodies, and is the key factor in the reserve energy, however, it is not the whole key.

The pH (measure of resistance) factor is number (2) and (3). The top number is the pH of the urine, and the bottom is the pH of the saliva. The numbers are not fractions.

The fourth number denotes the total salt measurement in micro numbers. This indicates whether the body retains too much salt or not enough. This is done through the process of elimination, which is a mathematical calculus equation.

The fifth number indicates the basic change of worn-out cells.

The sixth and seventh numbers cover or indicate many things. If these numbers are running too high it indicates that the proteins are turning to urea because of the high salt, and causes the heart to beat too hard, and the person is a candidate for a pectoris heart attack, also the person is too tired, or tensions are building up. If the worn-out cells are flushed out of the system in three days it does not turn to urea. Urea is undigested protein.

The iron, iodine, manganese, and arsenic are measured by the process of elimination.

These numbers show where, or in what area or areas, is the greatest loss of energy. The combined salt reading and the urea reading covers many reasons for the loss of energy in the body.

Practically all diseases start with mineral deficiency, which begins with the liver and has an effect on the other vital organs. As these vital organs begin to malfunction this affects the muscular action of the cranial nervous system, therefore it may make the symptom in a different location from the cause of the symptom.

Here again we run into energy cause and effect.

What the RBTI Test Will Not Do

By taking the tests consecutively certain dietary patterns can be determined. Then we can zero in on the health problem.

The test will not show gallstones because the gall bladder is so thin that when it stretches the energy is covering such a large area. However, there are other symptoms that show that there is a loss of energy in the digestion of foods because the bile has been in the gall bladder too long, and it is in a rather fermented state and therefore it upsets the digestive system. Then by deduction we can conclude that the gall bladder is not functioning normally. The stones receive energy but do not give it off.

Nausea, and soreness in the solar plexus area are symptoms of a gall bladder disturbance. (The solar plexus refers to the area where the ribs connect to the sternum bone. The solar plexus is at the bottom of the sternum.)

Where ulcers have been many times the gastric juices kept the ulcers so clean until the energy was dissipated over such a large area that we could not pick them up on the first test. After three or four days of tests they are very easy to pick up and locate.

The RBTI test does not pick up a broken blood vessel in which there is no loss of energy. It is just that the blood vessel is leaking a little bit of blood and no energy is lost because the tear in the blood vessel may be in the dermis of the skin or elsewhere in the body.

If a hernia is completely healed and there is no loss of energy it will not show. However, the hernia can still be there and cause trouble. If that hernia has inflammation, and there is a high amount of cell decay, it will pick up this hernia.

There must be a loss of energy in order for these equations to work.

People who have a buzzing or bells ringing, or humming, or a noise in their ears, it is actually a type of energy and can be picked up in the test. This is because the energy gives off a little bit of heat, and the heat waves are striking the anvil of the ear and this causes the buzzing, humming or noises in the ear which indicate an infection, or whatever the problem may be. This type of energy is easy to pick up even in very small amount.

These then are the weak points in the equation.

Chapter 10

RANGE A – E

In the RBTI equation there are five patterns that are away from perfect.

The perfect range has numbers so many decimals on each side of perfect. The ranges are A, B, C, D, and E. Range B is whole numbers away from range A. The C range is way out, it is a long way from range A. In range C the numbers are the farthest out above the ones in range A. That is, they are on the alkali side, or on the anionic side.

In the range D the numbers drop below perfect, or on the lower side of the A range.

In the E range the numbers are way out below the A and D range.

People in the E range are near death.

In these five ranges we have 2,600 differentials (i.e. we have 2600^{2600} different variations within the equation). To give a comparison of how great a number this is, here are some examples:

3^2 means 3 times itself 2 times (3 x 3 = 9)

3^4 means 3 times itself 4 times (3 x 3 x 3 x 3 = 81)

There are 80,000 standard drops of water in 4 liters (which is about one gallon) of water. That means there are about 23^{23} standard drops of water in all of the oceans (i.e. 23 times itself 23 times).

Now the RBTI equation has 2600^{2600} different variations. That means 2600 times itself 2600 times. That number is more than there are grains of sand or drops of water in the oceans. Do you realize that we are unique individuals and yet God knows each one of us?

It is possible for a person's body chemistry range to be in all five ranges. And this is where the differential comes in. We call these ZONES. It is in these zones that we measure named illnesses. For example, if one has a very low urinary sugar, it would be in range E, and yet the salt could be in range D, or C (or any range). Or, you could have pH in the C range and the liver pH in the D range (or any range), or vice versa. The same is true of the urea. This brings about what we call "zones." These zones can be graphed and in them the malady is located.

When one understands the numbers well enough, the frequency, the micronage, the milli-micronage and the milli-milli-micronage you can also see it in three dimensions, you also see the color, and you can draw a picture of what the problem is and where it is in the body with absolute accuracy.

This RBTI equation reading is not something simple and easy. It is being taught in nine seminars. We are covering a four-year postgraduate course in 45 days. We advise anyone to spend at least three months (six months is better), between each course so that it can be thoroughly learned or mastered. Every rule must be memorized and known as the multiplication table is known.

After working 38 years with this system in the three largest hospitals in one of our larger cities in Florida, I was not found wrong one time in doing this work for the doctors. Even surgery was performed when all the instruments showed nothing was wrong and the RBTI test was proven correct each time. Sometimes surgery was performed, when I had given the instruction that the patient's energy was too low to stand the operation, and would die on the operating table. There was not one exception.

With the RBTI test we have located everything from brain tumors to ingrown toenails without any physical examination other than the numbers, and later we did look at the blood vessels in the eyes to see if the body was cooperating.

I have challenged the scientists to prove this either to be accurate or inaccurate. At this point no one has accepted the challenge. Some have used smear tactics, trying to destroy it rather than to prove it right or wrong.

I am willing to take any case, without seeing the person, or having any case history, and in thirty minutes can tell by the numbers what the problem is. Hundreds of doctors have seen me do this.

We have taught nearly 1,000 people the basic principles of this course, and many of them are medical doctors and chiropractors. They are now using the RBTI tests and are amazed at the accuracy of this even though all nine seminars have not yet been taught. We have been teaching these seminars now for over two years.

My estimate is that 4% of all the people who take the seminars will master them and to actually see the picture that these numbers indicate.

Numbers do paint pictures, as they too are a language. An architect must see the picture in his mind before he can take the math that he knows and make a blueprint of the building, the bridge, or whatever he is going to build. He must see every number and every piece of material it takes to make whatever the blueprint calls for before an estimate of cost and time to build it can be completed. If an architect can do that with a building, why can't a biophysicist do it with a person's body? It can be done for I have done it.

I am not out to prove anything. I am out to make America the healthiest nation on earth. I am teaching principles that are in the Bible. When these principles are put into proper practice God will bless, God will honor, God will restore people to health.

Bad health is a result of disobedience to the laws God has made. Good health is obedience to the laws that God has made. I readily accepted the challenge to prove this when I taught the first seminar.

Chapter 11

HOW A NEW CELL IS DEVELOPED

A new cell does not develop by dividing. It forms by a process of ionization, the same way that a silver-plated, nickel-plated or chromium-plated process takes place.

The liver manufactures amino acids. (I have already explained how amino acids are formed and established, etc.). Then on the end of a nerve when it is cut it forms little branches. On each end of these little branches it forms a stolon which is a cation of energy which lodges with nowhere else to go. This process repeats itself and a cell comes into being. It does this on both sides of a cut, or bruise as an incision from an operation. This process is called knitting. If there is a mineral deficiency adhesions form and the flesh is hard. But if the mineral is sufficient then no adhesions form. If the mineral is plentiful there will not even be a scar left from the operation. It may be a little lighter in color, or maybe a bit of a marking there for a few months, or even a few years, but there will be no adhesions. All adhesions are caused by a mineral deficiency. Many times adhesions form within the body because of a mineral deficiency when there has not been any operation because the worn-out cells are swelled and there is not enough mineral to force them into the blood stream, or to break them loose from the nerve ending, and for a new stolon to form and a new cell to come into being.

Until the principle of the forming of a cell is understood, all diagnosis can be incorrect.

RBTI is not a system of diagnosis. It is a system of analysis. An analysis is something that is very accurate. It is very difficult to find a group of doctors that agree on any diagnosis, but if one million people did a RBTI test all of them would come up with the same answer and it would be accurate. (It is very easy to miss a diagnosis and to make a true analysis but it is impossible if your instruments and laboratory equipment are kept clean. However, humans being what they are, it is possible for a lack of training to misinterpret the analysis.) It practically eliminates human error because there are ways to prove whether or not the mathematics of the RBTI is accurate or not.

If you can think of a jigsaw picture with only nine different patterns of parts, but yet the parts of variable sizes, then you can imagine the shape and form of energy that makes up a human body, which is the most complicated of all the creatures God made.

If you would look at the brain under an ultraviolet ray microscope, it would look like a series of little rods. These small rods are not the brain at all, they lead to brain cells which are round, and at the end of each one of the rods is a very small round cell. They are in clusters. Each one of the clusters causes us to do certain things. In the left quadrant of the brain rests the involuntary system of the body which controls the vagus nerve system. The right quadrant of the brain is the one that controls the muscular action. The frontal part of the brain controls our thoughts, our knowledge and etc. Whenever we wish to recall something we are using the frontal part of the brain. If there is a potassium deficiency then electrical impulses do not travel in the desired paths to locate the information in the cluster in which it is stored. Therefore, we are absent-minded. But don't let that excite you at all because absent-mindedness is a sign of genius. Now, if it gets to the place where it is forgetfulness, it is rather serious because then delta cells are beginning to form and maybe the recall is gone forever.

Omega cells are premature aging and death has started. On a peach, apple or banana, if there is a rotten spot, as far as that fruit is concerned that is complete loss of reserve energy because it is decayed flesh of that fruit. Decayed flesh in our system is complete loss of reserve energy.

We have cancers that are benign, that means they are contained, and then we also have malignant cancers. However, a benign cancer may be malignant too, for it also may be growing very rapidly even though it is still benign. It is like a boil, which is a type of cancer. It has a rotten core in it, and that rotten flesh is cancer in any man's language. Any other rotten or decaying flesh is cancer, whether it is a pimple on the face, or a sore on the leg or arm, or anywhere in the system.

Our system should throw out a few hundred thousand delta cells every day and replace them with new cells. This is the process of life. Every cell in our body, from the brain to the tip end of the toe should be exchanged about every six months. The proof of that is the fingernail. They grow out about every six months. As long as this is happening, cells being replaced throughout the whole system, you are in perfect health, and in for a ripe old age. This is what the perfect equation denotes: How to grow old in good health.

Fifteen years ago I was called to the hospital early in the morning. I had been there nearly all day with doctors on various patients that had accumulated while I had been out of town. I had just gotten home on a Friday afternoon and did not wish to go back to the hospital when the telephone rang with the message that they had a case and for me to come in. A woman, age 35, had broken out in hives two hours earlier from the top of her head to the bottom of her feet. You couldn't put a pin down on her body anywhere but what there was a hive. She was about as red as a beet. The doctor asked if I had any suggestions to make without coming to the hospital. "No," I said, "I can only come to that answer by asking the lady some questions first."

The three doctors had examined her and one called me about the case.

I asked if the lady had had a family upset, family argument. No, she had not had any family upset.

Had she had any bad news from anywhere that upset her, like an illness or death in the family? No she hadn't had any of that.

How long after she ate did she have this attack? They said three hours.

Did she have cold hands or cold feet? No cold hands, no cold feet.

I then said that the hives were caused by a calcium imbalance, and the only thing that could have thrown her calciums out of balance so quickly was a heart attack. I asked how long it took the hives to come on. They answered, "15 to 20 minutes."

Did she have sharp-shooting pains in her stomach or chest? Yes, but it quickly went away.

How long from then until the hives broke out? From 20 to 30 minutes.

That meant the hives had broken out in a ten-minute period from the top of her head to the tip of her feet. Then I said, "See if she has not had a heart attack, and if she has not had a heart attack then I will come to the hospital. Call me back within an hour, let me know one way or the other."

One other question I had asked, Did she have deep wrinkles in her forehead? The answer was, "Yes." This indicates inward tension and potential of heart attack.

The inward cause of body chemistry will cause outward effects on the body. In the seminars we teach the students who take the numbers to be able to describe the person. They learn to describe the outward appearance. The numbers talk and give the inward appearance. (If your wife sent you to town to get a spool of number 40 or 50 thread, you would bring her back a thread one fiftieth of an inch in diameter. If you don't believe numbers talk just look at your bank statement the first of the month. Sometimes they even scream at you!) This is the result of cause and effect, the loss or gain of energy.

The doctors laughed at me for suggesting a heart attack.

They ran a test and found that she had had a heart attack, and that there was slight heart damage. I said, "You know what to do, and in the meantime I am going to suggest something for the hives. Put her on the lemon and water according to the weight, every hour, and water every half hour for ten hours a day. Also, give her asparagus twice a day, with some bread, and for the dinner meal only raw, leafy salads. I will be in to see her with you on Monday, and I think she will be able to go home."

I was in on Monday and went to see her with the doctors. She was feeling fine and all hives were gone. We gave her a diet to follow. She got along beautifully. This is an example of cause and effect.

Chapter 12

TESTING PROCEDURES

One thing about the Reams Biological Theory of Ionization tests is that they are done so quickly, in less than 30 minutes. When tests are made no case histories are required and no one has to be undressed. No questions are asked other than identification, name, address and telephone number for the purpose of getting in touch with the person as needed.

We do look at the eyes to know whether or not the body chemistry is cooperating, or not cooperating. In each eye, if the blood vessels were stretched out end to end, we are told there would be approximately 80 miles of blood vessels. In the sclera of the eye there are two groups of blood vessels. One running horizontal and one running perpendicular (upright). The perpendicular blood vessels indicates loss of delta cells that are not being washed out of the system from above the diaphragm and upper extremities. The horizontal blood vessel indicates that the loss of cells is below the diaphragm and the lower extremities of the body. The delta cells that are not being washed out of the system will eventually turn to urea and are indicating that a heart attack will be in three hours, days, or years, regardless if in the horizontal or perpendicular blood vessel eye dilation, unless something is done.

The sclera of each eye is divided into 60 parts, a total of 120 divisions for both eyes. By writing down the eye numbers of the sclera when it is checked off on our little checkerboard scale form, then the greater the amount of numbers or little boxes we have in the eye that has dilated blood vessels the greater the number the delta cells, or even omega cells, that should be thrown out of the body.

These blood vessels are so delicate until they are very easily stretched or expanded, and can also denote a vitamin C deficiency in the body. This is as much as the eye test tells us.

There are certain body conditions in which the body chemistry cannot accept vitamin C. Vitamin C is the element that knits the cells together. When women are pregnant and have a vitamin C deficiency they have stretch marks in their skin. Men with vitamin C deficiency may also have stretch marks. Stretch marks is always a sign of vitamin C deficiency. The RBTI testers do not have to examine a person to know whether or not there are stretch marks. Not the eye numbers, but the lab numbers tell whether or not there is a vitamin C deficiency.

Everyone who has a body chemistry with 4.5 milligrams of vitamin C per gram of blood has no type of disease. Vitamin C is the police department that keeps the diseases out of the system. There is a theory that vitamin C cannot be stored. *It can be and is stored in the system!*

One of the early signs of a vitamin C deficiency is a cold. A runny nose or sore throat is a very early sign of a cold. Also, swollen tonsils is a sign of vitamin C deficiency. Swollen tonsils are also a sign that the person is not drinking enough pure (distilled) water. I've never seen a case of tonsillitis, regardless of how badly the tonsils were swollen, that couldn't be brought down normally within a week with drinking the proper amount of distilled water systematically, plus taking vitamin C as soon as the body chemistry (pH above 6.40) was ready to accept vitamin C.

The taking of vitamin C does not mean that the body assimilates it at all. You may take vitamin C by the pound and the body not accept one milligram. One milligram is one thousandth of one gram, and it takes 28.4 grams to equal one ounce (1 ounce = 16 ounces = 1 pound). So you see that one milligram is a very, very small amount. Vitamin C has a low pH and if your body chemistry has a low pH it rejects vitamin C. There is one exception, when a cation is added to a cation, it will give off enough vitamin C energy to supply the required amount needed.

We can connect the deficiency in vitamin C to a direct cause of lack of calciums in the system. The vitamin C is deficient because of the lack of enough varieties of calciums in the system, plus the oxygen that we breathe, plus enough iron and iodine for the liver to function normally. So when we begin to study body chemistry we are going to run into chain reactions. One body action has a reaction upon another body action, and each time there is a reaction then our reserve energy level changes. Most medical doctors cannot find anything wrong with our bodies until the reserve energy drops below 30, sometimes to even 25, and less.

Whenever the principles of the RBTI are put into practice, all stretch marks can be eliminated in the skin. All cavities in the teeth can be stopped. All heart attacks can be predicted in a matter of minutes, to months, and even completely prevented.

Angina heart attack is not caused by cholesterol (a fatty substance). Cholesterol is the final effect that brought about the heart attack. The angina heart attack was brought about because of cholesterol, but the cholesterol formed in the blood vessels and in the veins and arteries because the body retained too much salt, causing the vessels to lose their ability to expand and contract so nature puts grease or oil in the blood vessels so that the blood can get away from the heart and back to the heart. Otherwise the person would die much sooner if the blood vessels dilated and the cholesterol did not form in the blood vessels. However, it is possible for enough cholesterol to form in the blood vessels, but a piece of this cholesterol can turn loose and plug up the heart, causing it to go into a spasm and cause a fatal heart attack, or damage to the heart. It may take years for the heart to repair itself unless steps are taken to supply the proper mineral needed to effect the healing.

When the doctor tells you that once a heart is damaged by a heart attack that it will not repair itself, he means that no drug will repair it. But diet will repair a heart that has been damaged by a cholesterol heart attack, or a pectoris heart attack if the body responds to a tailor-made diet. A pectoris heart attack is brought about because the system has not thrown out the delta cells, and the salt in the blood has broken the cell down to a urea.

Urea is in two forms—NITRATE NITROGEN and AMMONICAL NITROGEN.

These two nitrogens cause the heart to beat too hard. Not any faster, but much harder. When this reading of the urea (the last numbers in the equation) gets to 18 to 20, or 23, the person is very, very tired all the time because the heart is beating much too hard, causing the entire body to be tired. This is a very common occurrence with women because they use seven times more calciums a day than a man. Since the need for calciums is so much greater in females the weaknesses are magnified much more than in the male.

The lack of calciums causes the urea to rise higher and higher. If this urea rises above 24 to 27, this person is in the zone for a major heart attack. Above 27 anyone would be in the zone for a fatal heart attack, if it was a first reading. It shouldn't be that high even in a second reading.

Heart attacks, both angina and pectoris, can be blotted out within 30 days in the United States in a patient any time the people demand it. The quicker they demand it, the easier it will be.

Another thing that causes heart problems is acute indigestion. Fifty years ago acute indigestion was a very common disease. It generally could be cured in 30 minutes. But today there is no money in acute indigestion so everybody has "heart attacks" and it takes two or three weeks to get well, even though it is acute indigestion. It is very easy to tell if it is acute indigestion, or a heart attack. If acute indigestion, one has cold hands and cold feet. If it is a heart attack you may not have cold hands and cold feet. They are treated absolutely differently.

Acute indigestion is gas forming in the stomach and puts pressure on the heart. The heart has so much pressure on it that it cannot beat properly and goes to skipping. This causes chest pains very much like an angina or pectoris heart attack. Generally a pectoris heart attack is caused by the high urea in the blood. If it is coming along rather slowly it will cause pain in the left arm a long time before you have chest pains.

These are problems that can be prevented with diet, and be stopped forever. Someday a scientist will make an instrument that will be sold on the market so the people can use it and it will solve many of the common problems that are in the beginning of trouble or bad health for people.

The possibilities of what can be done with the Reams Biological Theory of Ionization equation is so great it is uncomprehendable by the human mind. We are truly marvelously made, and it is a matter of obedience to the rules that God has made, for us to be in good health.

Chapter 13

LOW BLOOD SUGAR

Approximately 80% of the people that are classified as having epilepsy, and on dilantin, or other drugs, are not epileptic. They have low blood sugar. The low blood sugar is causing seizures. There are no two seizures alike. No one by watching can tell the difference in whether it is an epileptic seizure or a low blood sugar seizure. The RBTI tests are the only tests that differentiate between an epileptic and a low blood sugar seizure.

A 27-year-old girl was brought in from Daytona Beach to be tested in Orlando, and had been on dilantin since she was two and one-half years old. After the tests were made I said to her, "So you think you are an epileptic?" The people who had brought her were her neighbors, said, "We know she is an epileptic because she had a seizure on the way over here."

"I'm not talking about seizure. I'm talking about the cause of seizures," I said.

I thought the lady had lied to me about her age because she looked 47 or 48 years old because she lived in fear of the next seizure. She was having as high as 15 seizures a day, and had to have someone with her around the clock. The dilantin and anything else they had tried was wearing out to where it no longer was effective.

I said to the lady, "I have news for you. You are not an epileptic, you are having low blood sugar seizures. You may or may not believe me. I am going to give you food which I want you to begin taking now. Take it until you eat supper tonight. Then I am going to give you a diet, and I want you to follow it. I want to see you in two weeks. If you do exactly what I tell you, you will not have any more seizures."

She did not believe me at all. Neither did the people who

were with her.

Now they really thought I was a quack.

I made up for her what can be made in most ordinary kitchens. It was just a matter of mixing juices together that would fit her body chemistry. It was something to keep her blood sugar from dropping too low.

In a week the neighbors who brought her called saying, "This girl has not had a seizure since starting on the diet."

"I didn't expect her to have any seizures. Bring her back in another week," I said.

They brought her back and I gave her a diet to follow and told her to come back in two months. In two months when she came back she looked her age. I then asked her to return in three months. When she came back she looked 20 years old! A year later she was able to get her driver's license. And a year later she was happily married. This is just one of the cases of a missed diagnosis, and the suffering this girl had gone through for nearly 25 years could have been eliminated if the RBTI tests had been used.

Another case was a 30-year-old man who was the head computer operator in a modern company in Orlando. The job was demanding so much of his time until he was about to lose his family. He was called on almost day and night and had no time for his young family. He had taken the contract to put a computer system into the hospital. After a few years he began having very serious migraine headaches, and they were getting worse. It was on Labor Day 1960 that he called me from the hospital.

"Do you have a headache now?" I asked.

"Yes," he said.

"Get someone to drive you to my lab. I can get you over the migraine headache in approximately 30 minutes."

I had to work him in between others needing dietary guidance on the Biblical health message. I had already been arrested two or three times, so it didn't make any difference if I got arrested again.

Those who instigated my arrest had forbidden the doctors to use my services. And the doctors who used my services, or wished to talk to me personally about their cases, came to me by night.

I told the girl in the office that the man was coming and to test him as soon as he came and to bring me the result of the test. When she brought the test I made up for the man four ounces of juice to sip slowly for 30 minutes. I gave it to the girl to give him, and didn't even go to see the man. I just said, "Give him this and tell him to sip it very slowly for 30 minutes, and his headache should stop when the 30 minutes are over." (This was not a drug. This was a food that can be found in most homes.) "Tell him I don't mean 28 minutes, and I don't mean 29 minutes, I mean sip it 30 minutes. But probably before the 30 minutes are over the headache will stop. And as soon as the headache stops you have him come and knock at the door." Nineteen or twenty minutes later he knocked at the door and I said, "Come in."

He came in and I said, "So you are the person who had the headache are you?"

"Yes," he said, "I want five gallons of that stuff."

"You don't need any of that any more at all," I said.

"What do you mean? This is the first thing that stopped my headaches. I was told just today to either get rid of the headaches or my company is going to get rid of me. I told them, 'You are the doctors, you get rid of it. I'll do anything you say.' They said, 'That's beside the point. You get rid of the headaches or we'll get rid of you.' And that is why I called you. Now then, you've got me rid of the headache with this for the first time and I want five gallons of it."

I said, "I'm sorry but you don't need five gallons of it. The numbers indicated your problem is tuna fish. You are allergic to tuna fish. Don't eat any more tuna fish and you won't have any more headaches."

"I like tuna fish! I eat it three or four times a week," he said.

"Well, that's what is causing the headaches. There may be something else you are allergic to. Come to see me in four to six weeks and let me know. But if you'll leave tuna fish alone you won't have any more headaches," I said.

In five weeks he called saying, "Mr. Reams, no more headaches. No more tuna fish. I'm fine."

Tuna fish in this man's system was creating a pancreatic enzyme reaction which was causing a displacement or confusion of the potassium and oxygen for the brain, and was short-circuiting before the blood got in his brain because of the particular poisoning this was to his system. What I'm saying is that allergies are a mineral deficiency. This is how accurate the RBTI tests are.

In this case you may ask, "How come if you eat tuna fish that has mineral in it that it created a mineral deficiency?"

The minerals in the fish prevented the minerals that he needed from becoming available. In other words, the digested fish created a catalyst, which prevented the potassium and oxygen from performing its duty in the brain. It was as plain as the nose on your face to fix. So much suffering can be prevented when you work on *cause* rather than *effect*.

Within the last year I had a call from a medical doctor who had done a test on a two-year-old baby. The first question I asked after he read the lab test numbers to me was: "Does this baby have seizures?"

"No," said the mother.

I then said, "The baby is in the zone for seizures. And since it does not have seizures, I don't see what the problem is."

"The baby has convulsions," the mother replied.

I said, "A convulsion and a seizure is the same thing. The problem with the baby is low blood sugar." I knew it was that and for the baby to have seizures, but if she wants to call it a convulsion then it is a convulsion. Most all babies that have convulsions, the problem is low blood sugar. The quickest way to bring a baby out of the convulsion is to place a little honey in its mouth and in one minute it will be out of the convulsion.

The modern tests that doctors do on blood chemistry for low blood sugar, and even the glucose tolerance tests, are very difficult to locate low blood sugar, and the blood sugar problems.

The glucose tolerance test is only at its best 50% accurate, and that misses it by a lot, too far for safety. There are those who have said it misses it 85% of the time.

The very first RBTI test will denote if there is a low blood sugar problem. For 23 hours and 55 minutes of a day the blood sugar may be perfectly normal. But for the next five minutes the blood sugar may drop too low, and the person feels like he is going to die, or may have a blackout. It is one of the worst feelings in the world. By the time he can get to a doctor, the doctor can find nothing wrong because the adrenalin glands have started to flow and the blood sugar is back to normal, and the doctor says, "It is all in your head."

It was all in the head, but the doctor didn't find out the cause for it. In low blood sugar the blood does not take enough oxygen to the brain, and it was "in the head."

I have had people with low blood sugar that would go into coma at any time of the day or night, and the doctors knew nothing to do about it. In a matter of two or three weeks with an individually-made diet we would have them completely free of the low blood sugar problem.

Low blood sugar is caused by the pancreas manufacturing too much insulin. There are foods that cause the pancreas to manufacture and flush too much insulin. So the doctors take patients off of all sweets and carbohydrates because they have found some whose pancreas flushes because they use certain sweeteners or carbohydrates, and they put them on an all-protein diet, which raises the urea, causes them to be more tired, puts them in the zone for pectoris heart attack, and, therefore, "robbing Peter to pay Paul." The patient won't die at the moment but probably will die much earlier with pectoris heart attack.

This type of person must be in a retreat and each day have a given diet, and each day the medical doctor on the staff serving that patient can tell them by the diet they had the day before which carbohydrate has caused the pancreas to flush, or which carbohydrate caused it not to flush. Then the patient is taken off those carbohydrates that cause the pancreas to flush and they no longer have low blood sugar. It is just as simple as that. But these tests are the only tests that can detect every stage of high and low blood sugar, and *are* accurate.

In the low blood sugar range there are people who have other problems besides the low blood sugar problem. They may have the low blood sugar and may have a very high urea. The diet for the very high urea, and the low blood sugar are opposite. There is no way to handle this kind of problem without having the person in a place where they can be tested two or three times a day so that blackouts or seizures can be prevented. It only takes about two weeks to correct this kind of a condition. I wish it were possible for us to give everyone a diet so no one would need to go to a retreat, but this is not possible because of the necessity of running a test many times a day.

The American Medical Association often complains about the prices that are charged in the retreats for services that we give on diet. The prices are about the same as staying in an A-class hotel. This price is the total of all tests, dietary aids, and diets. The hospital rates generally start at $75 a day for room only, besides the doctor fee and other tests that are made. In one Los Angeles, California, hospital that I was in this past year for 12 days, my room rate was $350 a day. Now, who is giving the rip-off, the medical doctors in the hospitals, or we in the retreat with the health message as written in the Bible? We are not running hospitals. We are not running drug centers. We are running seminars in which we teach the health message as written in the Bible. We are not treating diseases. We are teaching people how to live without drugs and to live healthy.

Many people are coming to spas and retreats because drugs have failed to get them well. Operations have failed to get them well. Many of those coming are young people, some just children. As a means of last resort they search out any place where they have hope, where they can get encouragement. They are disenchanted, disappointed, brokenhearted over brainwashed ideas of what drugs can do.

The drug establishment is trying to make people stay on drugs even when many medical doctors know that no drug can heal.

Medicine is the practice of drugs, it is not a food. In fact, many doctors pooh-pooh the idea of food having anything to do with a person being in the best of health. Therefore, when I am accused of practicing medicine without a license on a subject that is pooh-poohed by the drug establishment, it proves that I am not practicing medicine without a license. I am practicing *drugless* therapy, and medicine is the practice of drug therapy.

If you wish to get well on drugs, I have no objection.

If you wish to get well by the use of God's plan of a better diet, I have no objection; or the use of both drugs and food, but I do maintain that this is America and that you have a freedom of choice to choose the way and to whom you should go with your health problems. You should have the right to choose if you are carried to the cemetery young, or live in the best of health to a ripe old age. Anything less than this is tyranny.

Chapter 14

OBESITY

Obesity, or not obesity, that is the question!

There are many causes for weight problems, and only about 2% of overweight people are obese. I would venture to say that 95% of the 2% are ill. There is a reason for them to be obese.

A 28-year-old woman came to my office who weighed 550 pounds. However, she was 6' 4" and had a big frame. It was all we could do to get her through the door. She was the biggest person for whom I had ever written a diet. It took two scales to weigh her. She had six children. Two years prior to coming to us she had been put on the birth-control pill, and immediately began to gain weight. For the previous two months before she came in she had gained 55 pounds each month. She had been told she would be dead within 90 days because there was no way to stop this weight gain. They owned an ornamental nursery, and she was a very hard worker.

The first thing I did after the test was to suggest she stop taking the birth-control pill. Second, to take her off of all wheat and milk products. I then furnished her with a refractometer to measure the carbohydrates in the urine so she could keep them between 1 and 2 brix (the measurement of sugar per pound of volume) on the refracto dial. I told her to eat nothing but vegetables, lots and lots of raw salads, and no meats at all until I gave her the word to eat meats. She was to drink raw vegetable juices made with a juicer. She didn't have a juicer, she had to buy one. I wanted her to come back in two weeks.

After two weeks she had lost 12 pounds. I then put her in bed for two weeks on nothing but lemon water and distilled water (four ounces lemon water every hour for 12 hours a day, and four ounces of distilled water every half hour for 12 hours a day). All she was to do was to go to the bathroom. No work. No nothing.

The reason for not being on the lemon water for the first 14 days was because the urea was so dangerously high that if she had gone into a withdrawal it could have caused a heart attack. The diet I gave her brought the urea back down out of the zone for a pectoris heart attack.

Then she was put on a very, very light diet for another two weeks. I was to have a report every other day. If she got weak or hungry I was to be notified. The thing about it was I wanted her to rest. If she had tried to work then the lemon water and the water would not have taken effect. She would have been burning up the very energy she needed in order to restore her body chemistry so that it would rebuild itself.

The next 14 days she was to have a very, very light diet with a very small amount of food and continue on the lemon and water. She said she lost no strength, that she felt wonderful. By the end of the fourth week it was very difficult to keep her down. By this time she had lost 60 pounds of weight. I restricted her work to one-half day in the ornamental nursery. It was to be very, very light work. She was to continue on the lemon and water and a very light diet, along with lots of vegetable juices, very low carbohydrates, a little fruit for breakfast and no more throughout the day. She continued to lose 50 pounds a month until her weight dropped to 260 pounds, which was about normal for her big size. To this day she is still that size. This was in 1968.

The refractometer I had given her is an instrument to measure the total carbohydrates in the urine. She kept her urinary sugar between 1 and 2. She had to keep it there and it was very difficult to do because her own fats turned to sugar. But she did it and is the picture of health today. She can pick up and carry 300 to 400 pounds.

This is one case where the birth control pill caused overweight. It caused the pancreas to stop manufacturing thyroxine. The pancreas takes the glycogen from the liver and manufactures thyroxine out of it. It goes to the thyroid gland which adds potassium to it, and the potassium then forms with the thyroxine and forms a substance similar to grandma's old-fashioned soap, which dissolves the excessive oils in our body and is a weight-control factor. It is the governor for weight. The birth control pill upset the thyroxine manufacturing in the pancreas and therefore it didn't manufacture enough and the lady just kept putting on weight. This is one great cause of overweight.

You can go on all the calorie-counting diets you want and if this is the problem you will starve to death and still won't lose any weight.

There are other factors involved also, but this is a very serious one that really upsets the weight. If the pancreas should manufacture too much thyroxine and the thyroid adds potassium to it, then you would have trouble putting on weight. If it doesn't manufacture enough you will have too much weight. Doctors give thyroid extract with the potassium for the thyroid gland, which is the worn-out cell of the thyroid gland. The thyroxine is the cell that is not worn-out. Which makes all the difference in the world in the results you get.

You may have a case wherein the pancreas is manufacturing the right amount of thyroxine, but the body does not have enough potassium, therefore, you can be over or under weight from that problem. But probably it would be overweight. But it may be either one. The RBTI test tells us definitely whether this is the cause of overweight, or whether it is something else.

The reason the liver is not manufacturing enough glycogen is that there are not enough calciums of the right kind present. Therefore the liver doesn't have the material to make the amino acids that the pancreas needs to manufacture the thyroxine to control the body weight. For people who are too skinny the thyroid manufactures too much thyroxine. For the overweight it doesn't manufacture enough thyroxine. This is easily remedied by the correct diet.

Overweight can be a genetic cause. You cannot do very much about the weight if it is a genetic thing because the genes at the time of conception determine your weight throughout your life. Many people will be thin until they reach 50 years of age, or varying ages, and then it's a matter of looking like senior citizens. Generally people with big bones, broad shoulders, big head, and large feet have more weight than they want. But if it is a genetic normal weight there isn't anything you can do about it.

If your body is in any range outside of A of the RBTI test

you can take a diet and bring your body back to range A, and then your body weight will drop to its genetic normal. However, it may not be the weight you desire. You will generally look in weight like some of your aunts, uncles, or grandparents, but you'll wish you didn't.

Another cause of overweight is a lack of calciums in the diet, which can cause extreme nervousness. These people eat, eat and eat. This is not obesity, this is a nervous condition. They eat in order to overcome the nervousness. In eating they are trying to get more calciums and yet the body doesn't pick it up. Nature is craving food like an alcoholic craves his drink, as a person who is on pot craves dope. Hence the gaining of weight. Also, we find that by eating it helps to control temper. If these persons don't eat they become very irritable, and very difficult to live with.

Other causes of overweight:

- People who eat too much fats, foods that are too rich, too much carbohydrates, and especially those who eat them late at night (especially carbohydrates) when the body already has a sufficient amount.
- Eating too much potato and bread. The starch in the potato can be converted very easily into sugar by the liver. A very poor choice of foods can be the cause of overweight.
- Not drinking enough distilled water, to regulate the foods.
- Slow digestion of food. It digests so slowly until the body chemistry is starving all the time. They eat all they can hold, feel stuffed.
- Skip breakfast, have a brunch. At noon they have a lunch, and in the evening eat a big, big meal. They get up feeling miserable in the morning. They feel like something the cat has drug in, and they can't do one-tenth of the work that they ought to be doing. They are trying to hold a job. They are upset and irritable. The brunch does not even digest. They have their lips poked out for someone to step on, and generally someone accommodates them before the day is over. Their favorite song is that "Nobody Knows the Trouble I've Seen." This too can cause overweight: just plain frustration.

These are factors that need to be determined before we can pinpoint the cause and effect of overweight.

Weight is very easy to control. Anyone who will keep the carbohydrate sugar level between 1 and 2 of the RBTI test, eat a great variety of foods, especially vegetables, and cut down on meats (way down on all kind of meats, and do away with the eating of unclean meats), I have never seen one person whose weight wouldn't drop to its genetic level. (By unclean meat I am speaking of those referred to in Leviticus 11: pork and all its products, including ham, bacon, sausage, pork chops, spare ribs, pork roast, salt bacon, cracklings—any part of a hog is unclean to man. All the shell fish, including shrimp, tuna, mackerel, catfish, oysters, lobster, clams, scallops. Any fish without fins and scales is unclean for human consumption. Then there is also the rabbit, guinea pig, crow, buzzard, bald eagle, muskrat, and snakes.) The unclean meats digest in a period of an average of three hours, causing more energy to be released than a body can consume. The clean meats digest in an average of 18 hours, giving the person a chance to use the energy as it is made available. The same principle is putting high-test gasoline in the car that is not manufactured for it. The car will run better if regular gasoline is used. The high-test gasoline will burn up the pistons and the valves. The steel and the iron is not tempered for the heat of the high-test gasoline.

I am against the abuse of foods. Use temperance in all things. I am not against fried chicken, but there are people who have it every day. I am not against coffee, but there are people who drink 20-30 cups a day. Coffee many times can be an asset to people with arthritis or bladder or urinary problems, even weight problems. Coffee with chickory is good for people with low calciums and have nervous disorders. Coffee should be weak enough for you to see the bottom of the cup for it to be a diuretic. Watermelon juice can be a very fine diuretic, providing that you are not fat, or do not have high blood sugar. Some people can take watermelon juice and have high blood sugar without raising it, but some cannot.

There are conditions in which people have high blood sugar and also urinary problems. These kind of conditions require opposite diets. Even in a retreat under the very best of controlled diet, it is sometimes very difficult to make a diet that will show immediate improvements for the person to recognize the improvement. However, in a retreat the testers and the doctors who are supervising can tell from one test to another whether the person is improving or not. No one is kept in a retreat whose body does not respond to diet. If the body chemistry does not respond in seven days they are asked to leave because it simply means that the RBTI is not effective for their body chemistry. There is less than one percent of all people tested whose body chemistry will not respond to these tests.

There are people who come too late to the retreat. But if we can keep a person living for 30 days, and their body chemistry continues to improve, and they follow through on the diet after they go home, take the test as instructed, as far as we know, they all get well.

There are people who wish to go back into their old way of living and break all the rules of health, and commit suicide.

One man, who was only in his late thirties, had acute emphysema. He began the RBTI diet in the early summer. It was not necessary for him to be in a retreat. He could do his food chemistry at home. By November there was no sign of emphysema in his lungs. I personally instructed him not to go out in inclement weather. He had a poultry farm. I told him it wouldn't make any difference if every chicken he had died, "Don't you go out in bad weather." All of the new cells in the lungs were so young, his lungs were about as delicate as a newborn baby. They had not yet built up the reserve and strength of adult cells. I warned, "If you do go out in inclement weather you could take double pneumonia and you could be dead in 48 hours. DON'T GO OUT!"

He came the first of December for the test and I warned him again. One cold, wet December day they had some baby chicks come in, it was raining, the wind was blowing, and one of the employees didn't come and he went out to tend to the baby chicks. That night he took pneumonia and 12 hours later was dead. This is suicide. We can't help that. An autopsy was done and the doctors were surprised that there was no emphysema in either lung.

Overweight is not the cause of death for anyone. Death results because the body chemistry is in the wrong range. More people underweight, or who have normal weight, die in proportion to overweight people.

Chapter 15

FADS OR FACTS

I am often asked, "Is this food or that food good for you?" Many times people think I just do not want to tell them. They ask the RBTI testers if this or that food is good or bad for them, and the testers give the same answer: "I do not know whether it is good or bad for you because we do not have scientific facts to back up our answers. Until we have the RBTI test made we do not know whether something is good or bad for anyone, unless it is a no-no, such as intemperance on anything, or the unclean meats, or the failure to rest one day in the week, or that we make a garbage can out of our stomachs, or that we are not drinking enough distilled water."

There are some rules that can be followed which will greatly aid most adults, or even children, in some respect:

- Eat the heavy carbohydrates early in the morning, unless you are a diabetic and are instructed differently.
- Eat the large meal, the heavy meal, in the middle of the day.
- Eat a very light meal in the evening.
- Do not nibble while you are watching television at night because you will likely find the night sleepless. The lack of sleep has never killed anyone, but worrying about sleep does kill.
- In the American diet most adults eat too much meat and not enough vegetables. Everyone should have a raw (green, leafy) salad before the main course, at least once a day, twice a day is better, if they wish to stay healthy. Salad dressings may be used if desired.
- People should drink distilled water systematically, and should keep their body in rhythm. It is very important to have

meals on time, or within a half hour one way or the other. There should be a time for a bath each day, and a definite time for your bathroom duties. It is a very good idea to brush your teeth after each meal. In doing this your body gets accustomed to a rhythm, and it is ready for food at a given time. But if it never knows when you are going to take food it throws the body chemistry into confusion, which results in a loss of energy because the body is unprepared for the food that you are eating. Occasionally this is all right, but whenever it is kept up day after day, week in and week out, month in and month out, year in and year out, anyone is asking for trouble. Even for men it is a good thing to have a definite time to shave. It makes a world of difference in your whole day to keep your body rhythm as near perfect as possible. An occasional break in this rhythm will not do any harm. An occasional big meal in the evening will not do any harm, but a continual breaking of these rules will affect your health.

What foods are nourishing to one person will be poison to another. There are genetic reasons why this is true. There is no one food that I know of that agrees with everyone.

Some advocate there are combinations of foods that one should not eat, such as acids with alkaline, milk with acid fruit, etc. This is taboo. Energy is created by putting acids against alkalines.

In 1973 I recommended a diet for two ladies in the retreat who were in their seventies. One of them had worked all the days of her life, the other kept house for her. The one who kept house for the other worked tirelessly for her sister. They came to the retreat for a tune-up. They went on a 3-day lemon and water fast. There was no sweetening in the lemon. They went through a mild withdrawal. On the fourth day the one who had worked out decided that she could not eat the poached egg and toast for breakfast. The cook could not change the menu without a written order from me. The lady said, "Mr. Reams, I am allergic to eggs, I cannot eat eggs. I haven't eaten eggs in forty years. Now you have given me eggs for breakfast."

I got out her record of the past few days, looked it over and said, "I see no reason why you shouldn't eat eggs. I want you to eat eggs for breakfast."

She said, "They will make me sick."

I said, "Good. This is the place, if eggs will make you sick, get sick. This is why you are here." I knew she was not allergic to eggs.

She *did* eat the egg and toast. It is very important to know the next day what the person ate the day before in order to know what foods would give them the most energy in the shortest length of time. That afternoon I saw her and asked, "Well, are you sick?"

"No," she replied, "I'm not sick. Those were the best eggs I've eaten in forty years." She wasn't allergic to them at all. It was all in her head. She had an idea eggs would make her sick so had not eaten them, and consequently there are minerals in eggs that she needed and were missing in her body chemistry.

A few days later there was a half grapefruit served for breakfast and a glass of skimmed milk (non-fat). There again she refused to eat the food that was set before her. This time she came to me and was quite excited. "Mr. Reams," she said, "you are supposed to know about foods and this is supposed to be a health retreat. Don't you know that you can't mix milk and orange or grapefruit? One is alkaline and one is acid and it will make you sick. Don't you know you can't do that?"

"No," I said, "I don't know that. Nevertheless, this is what you are going to eat this morning for breakfast."

She said, "It will make me sick."

I replied, "Great, go ahead, get sick. Let's find out whether we are on fads or facts. Here we go by the numbers. You go eat your grapefruit, and drink your milk. Then we'll see if you get sick. Please let me know if you get sick."

I didn't see her again until the next morning. When I asked, "Did the milk and grapefruit make you sick?" she replied, "Didn't hurt me at all. I felt wonderful all day. In fact, I feel better than I have in years."

Many times people have fads that they think are the healthy way to live, and it is about the worst thing they could do.

There is only one combination that I know of that I do not recommend and that is mixing pears with celery. It won't do any harm, but the juices of each neutralizes the other and not enough energy is given off to do any good. It is almost like water when you put these two juices together.

Some who come to the retreat are surprised and say, "We thought you ran a health retreat, and you serve coffee!"

When people come who drink 12 to 30 cups of coffee a day you cannot take them off of coffee instantly. It may cause them to have a heart attack. We take them off of coffee slowly. Coffee is more dangerous to stop quickly than cigarettes. I have never known anyone to be harmed by quitting cigarettes instantly, even if they were smoking three packs a day. But I have known people to die that quit coffee instantly when they were drinking 20 cups or more a day. And they did it on their own, it was not on any program that I recommended. They just simply decided to quit. They became so religious in time that they quit coffee, and all of a sudden the heart simply stopped. Caffeine is a stimulant for the heart, and if you are inebriated by caffeine then please do not quit coffee quickly. Take a week or two to quit if you are drinking over a dozen cups a day, and a week if you are drinking less than a dozen. Then drink coffee only as you need it. One cup of weak coffee (you might call it Brazilian Tea—you can see the bottom of the cup) is very good for people over forty.

"Throw away the frying pan," is a half truth that is put out as fact. This came about over 100 years ago when the only thing they had to fry foods in was hog lard, or shortening. There is no such thing as vegetable shortening. It is either vegetable oil or hog lard. Hog lard is shortening, and if the label reads VEGETABLE SHORTENING, it simply means some vegetable oil and a lot of hog lard. So don't be fooled by reading the label (they forgot to put in a comma).

Even Crisco, which is supposed to be 100% all vegetable has "shortening" written on the can, simply because some people don't know what shortening is, and therefore that word should be removed from the label. Crisco is pure vegetable oil.

We need to use more oils and more fried foods than we have been taught to use, but use the better grade of vegetable oil. If there were more oil (corn oil) used today in our cooking, especially in the steaming of vegetables, and in casseroles there would be less constipation problems. I highly recommend the use of corn oil, or a high quality olive oil, for frying or cooking because it is easier to digest than any other oil. A high quality olive oil is tasteless, and if it tastes strong it should not be used. Corn oil does not have the amount of preservatives in it to hurt anything.

The vegetable oil that requires a very low solvent, such as corn oil and olive oil is very easily digested. In the liver they form gelatin, which prevents the undigested foods from sticking to the lining of the intestines and makes elimination easy. Now don't expect the first time you use vegetable oil to get a perfect elimination. This is something that takes weeks and months to get your system to really synchronize to manufacture a sufficient amount of gelatin to afford the body to give a good, clean elimination. If more vegetable oils were used in cooking today there would be much less diverticulitis.

There are other vegetable oils like Wesson Oil (peanut oil), Sesame Oil, that are very good if you are very healthy, but if you're not healthy you should use oils using the least amount of solvent possible. Safflower oil is one of the more difficult oils to digest; however, a little bit of safflower mixed with lemon juice makes a very good salad oil dressing.

In the cooking of most meats I suggest using one tablespoon of Manichewitz medium dry grape wine to each pound of beef, lamb, or venison, and one teaspoon in cooking chicken. Also, in cooking chicken use some thyme when it comes to a boil (or use the thyme in the water in which the blood is soaked out of the chicken before cooking). Thyme takes away the animal taste.

We recommend chicken and fish once or twice a week; beef and lamb once a week for those who are meat eaters.

Meat should be soaked in salt water 12 hours (using twice the amount of salt that you'd use if you were salting the meat for cooking) to get the blood out, and then to soak it in fresh water 24 hours. If meat is good, it will be better without the blood, and if it's not good, taking the blood out won't make it any worse. Soaking will not toughen the meat. Adolph's Papaya Meat Tenderizer is excellent to use. Papain from the papaya is a white powder (an enzyme in the juice of the green fruit of the papaya and apparently intermittent in action between pepsin and trypsin) and tenderizes the meat, making it much more digestible.

The only difference in the chemistry of raw and cooked

meat is that the cooked meat gives up the moisture, and as it gives up the moisture the tissues become tenderized. Meat should be cooked until well done. Medium rare and rare meat should never be eaten if you want to be healthy. They are much more difficult to digest.

There is the idea that fertile eggs are better for you than eggs that are not fertile. There is only one cell difference in the fertile egg. For it to make any difference you would have to use at least 2,400 eggs at one breakfast.

These facts can be substantiated by laboratory tests and only those facts that can be substantiated by laboratory tests are of any lasting value. It is very important to study foods and food chemistry, and also to read the labels when you buy foods. However, the labels are written by chemists for chemists, and most housewives and people who buy foods are not chemists, and, therefore, they should use dictionaries more to find the meaning of the words on the labels.

Certain foods slow down digestion in most people. Such as cheese, which not only slows down the digestion, but also is constipating. However, if you have loose bowels, cheese is an excellent food. If you have a constipation problem and poor elimination, you should go very light on cheeses.

One should not eat over two slices of bread a day. If you eat more bread than that you are threatening the nutrients and minerals found in other foods. It should be toasted for best nutrients, and it makes it easier to digest.

There are people who need a little wine, or Zest Tonic with their food. Not before or after, but with their food. People who have cold hands and cold feet all the time should have a little white wine with their food. Zest Tonic is the nearest product ever made to the exact enzymes that are made by the pancreas and is better than wine. Everyone has a built-in whiskey still, or gin factory, so to speak. It turns the excessive sugars into alcohol, and it is the alcohol that determines our body temperature. I'm not speaking of fever, I'm speaking of a normal body temperature (98.6° F). Generally people over 40 years of age are troubled with cold hands and cold feet, or are cold all over. This is because their pancreas is not manufacturing alcohol.

It is not the use of alcohol that is bad, but it is the abuse of the use of alcohol, and also the using of the wrong type of alcohol. The RBTI tests will indicate which, and how much, of the four kinds of alcohol is the best for your body chemistry. It is recommended that you use 1 teaspoon of Zest Tonic in four ounces of grape juice after meals and sip it (sit down and do absolutely nothing) slowly for 30 minutes, or until your feet and hands become warm. After a few weeks you will notice that each time you do this there is a mild headache that will follow. Do not quit using the Zest Tonic because of the mild headache; however, cut the amount in half and when a ½ teaspoon full, sipped for 30 minutes, after dinner and supper, if it gives a headache, cut it down to ¼ teaspoon, and when that gives you a headache, use only a drop or two, and when that gives you a headache discontinue it completely. The pancreas will again be working perfectly.

Some people go to a doctor and he says, "No salt. I want you to have a bland diet," so people buy salt substitutes, which is just as bad as salt itself because salt substitutes simply means it is not sodium chloride (which is ordinary table salt [NaCl]). Other salts can be magnesium chloride, iron chloride, ammonia chloride, potassium chloride, carbon chloride, etc.

There are salts that do not contain chlorides in them whatever. Some are ammonia salts, nitrogen salts, and a number of carbonate salts.

All of the salts cause the blood vessels to dilate, and also the intestines to form diverticulos areas in them, and generally upsets the body. Some people, before they taste the food, empty a salt shaker and yet their salts are perfectly normal in the tests because their body chemistry is perfect enough that it throws out the salt and does not retain it. There are other people who do not use any salt, or salt substitutes, and yet their salt is dangerously high. Sometimes high enough to cause or bring about an angina heart attack. *It is not how much salt you use, it's how much salt your body retains.*

Sea salt is the worst of all salts for people to use. In many people it causes a dry, hot tongue and it dries up the saliva glands and the mouth is always dry because sea salt contains seven kinds of salt and only the healthiest log rollers (or a man doing heavy muscular work) can use it, and they should use it sparingly. Just because it comes from the sea does not mean that it is better.

Some people are so afraid of food additives until the fear of the food additive does them more harm than the additive itself. "Perfect love casts out all fear," and there is nothing in the life to fear but fear itself. It has been said that "a coward dies many deaths but a brave man only dies once." There has never been a time in the history of American foods that they are as safe as now. These additives make our foods safer and have a more lasting result than all the vaccines that some doctors give.

There are foods that do have too much additives in them. One is ice cream. It has many additives, it won't even melt. It's a smooth, velvety, sweet flavor, but it has very little food value. It will rot instead of turning to a whey. Any ice cream that will not melt, or will not turn to a whey, is unsafe to use for food, as it does not have much nutrient in it. Good ice cream will melt into a milkshake, and there are some of these now on the market.

Some orange juices are diluted with water and have a preservative that makes it dangerous to drink for people with a high cholesterol or a high urea. It is always safer to drink fresh orange juice.

I have a very high respect and highly recommend Ocean Spray Cranberry Juice and Welch's Unsweetened Grape Juice. If these companies keep their products in the future as well as now they will be great products in years to come. The unsweetened grape juice has the natural grape sugars in it. I also have great respect for Dole and DelMonte Pineapple Juice. Pineapple, unsweetened, whether in slices or crushed, is very rich in minerals. Most everyone should eat a little pineapple once or more a week (unless they are allergic to it). Always use the fresh fruit whenever possible. When buying, always buy the fruit that feels solid and is the heaviest.

Be temperate in all things and use a great variety of foods. It's much easier to maintain a higher reserve energy rating by following these principles.

Chapter 16

MIN-COL

While in college I became aware of the great mineral deficiencies in our foods.

In 1934 I discovered the best of the minerals that is still on the market today, Min-Col. This is a derivative from bone meal which is very effective in rebuilding bone structure, etc.

Min-Col is the essence of bone meal. There is approximately 60 pounds of this substance in a ton of bone meal, but man has never learned to get over three or four pounds out of a ton. It is a bit expensive. There are 100 capsules in a bottle, so 3½ bottles would be enough for a person for a year. Therefore you have spent approximately $50 (unless during pregnancy), besides no cavities, no dental bills and strong teeth and strong bones. What you spend for these capsules decreases what you would have to spend with orthopedic surgeons and dentists.

There is no other product on the market today that I know of that even remotely equals the high quality and standard of Min-Col. There are some counterfeit substances on the market of very low quality (have not been proven over a period of 45 years). I raised six children with Min-Col with no cavities in their teeth and no receding gums, (plus a good diet) and no days missed from school because of illness.

Six-month-old babies should take ½ of one Min-Col capsule daily. Let the mother break it open and put it on the foods. This will cause the child to have strong teeth. They will not have cavities or lose their teeth too early. At four years of age they should take one capsule a day for 96 years, and if they live to be older than that, repeat the dose.

Females should take two capsules twice a day during

pregnancy, assuming they have been taking one capsule a day previously. After the baby is weaned, then go back to one capsule a day.

When fingernails tear easily, are rigid, thin, have rough edges or have white spots, this is a sign of a need of Min-Col.

When the calciums drop too low people become irritable and nervous. If you take the wrong calciums, or you are not taking enough of the right kind of calciums, then problems will result. We need to keep in mind to be sure to use the right calciums in the right amount at the right time. By right amounts you have to consider age, height, weight and diet.

Healthy people's diet should be 30 to 35% roughage, or cellulose (even in animal feed this should be the same amount). If you will use 35% cellulose fiber in your diet, then you will not have any digestive problems such as constipation. After you have become ill and you have problems in the colon you will find that the liver bile is too dilute to break down the cellulose enough for it to be effective, and the lower the reserve energy, the weaker the digestive juices and the less minerals you get out of the food. You are caught in a cycle of a chain reaction when everything you do is wrong as far as correcting your diet. Until you get your body back into range A of the RBTI test everything you eat is against you. So you need help in getting your body chemistry back into range A. This can be done under supervision in less than a week, generally in three or four days. It does not mean that you are well the first day the body chemistry is moved back into range A. It indicates that nature is repairing the damage as rapidly as possible.

One thing to remember about the numbers is that a change in one number indicates a change in all numbers. Something like the Ten Commandments, if you break one, you break all.

It is almost impossible to fool these numbers because of cause and effect. "Well, my numbers change quite often," you say.

The lower your reserve energy the more effect your diet has on your numbers, but if you have a high reserve energy you can eat most anything and it won't change the numbers. The very fact that your numbers are jumping around means that you have a low reserve energy.

Avocados and mangos are very high in minerals. Avocados

are as rich in proteins and minerals as is a beef steak. Mangos are a very good fruit and rich in minerals, but if it is very ripe you need to get into a bathtub to eat it because it gets all over you. Some people are allergic to mangos or avocados. People having this problem should not use them. Remember, there is hardly any food that somebody is not allergic to. Where can you find mangos? You will find that mangos (man goes) wherever woman goes.

People with ulcers of the stomach, high delta cells in the stomach, colitis, colon pockets, hemorrhoid condition, and inflamed pancreas, and inflamed appendix, should not eat nuts or popcorn at all. It doesn't matter how well you chew these foods, little particles collect in the decayed spots or in the ulcerated area, or in the colitis where irritated in the intestinal tract, and can cause soreness, protrusion, swelling, and a lot more suffering than is necessary.

Refrain from the use of nuts, with the exception of coconut. People with digestive problems such as just mentioned can use Aloe Vera Gel (one tablespoon twice a day, between meals, taken in carbonated drinks such as Collins Mixer, 7-Up, etc.). Aloe Vera Gel helps shrink hemorrhoids. It is one of the finest healing agents for digestive problems.

It is good to learn to tell your tongue what you are going to eat, whether it considers it good for you or not. Evil spirits can get in the body of even saints at times if they do not eat the proper diet. The evil spirit knows that the weaker the saint becomes physically the less faith he will have, and the less capable he is of making up his mind and the less works he will have. So it is the evil spirit's job to decrease the strength so the works won't increase. The demon is sent into your body for the sole purpose of killing you, and when he kills you he does not die, he goes into someone else and does the same thing. Evil spirits in people does not mean that they are lost by any means. It is something like the rat being in a hole in a happy home. He is an undesirable visitor. Evil spirits demand the foods that they like, and they will demand it until the person is dead, eating the very foods that are killing the person and making them hate the foods that the person needs to recover his health. It says in the Bible that Jesus cast out evil spirits and the people became well immediately. This is a miracle. However, if you refuse to eat the foods that the evil spirit demands, you are resisting him as is written in James 4:7, and he will leave your body because unless you obey him, he will not stay. For it is in essence whether or not we control our taste buds (or tongue), or they control us.

Chapter 17

CASE HISTORIES

There are three classes of people:
Those not sick who want to be.
Those very sick and they know it.
Those very sick and don't know it.

This last group is a sad group, and this is particularly true of athletes. At the retreats when moving the body chemistry of young men from range C to range A, they will go through severe withdrawal, probably vomiting like an alcoholic would go through when he is sobering up. I have these athletes to tell me, "Mr. Reams, if I ever get over this, I'm going to whip you, soon as I get well. I've never been this sick in my life." They feel so good when they get over the body chemistry change they forget to whip me. The body chemistry changes really happen and no drug can cause the body to change from one range to another, only minerals can do this.

Let us take the case of a female, age 50, weight 130, height 5'5", with the following numbers:

$$.08 \qquad \frac{5.2}{5.4} \qquad 18C \qquad 4M \qquad \frac{4}{6}$$

Here we have a person with low urinary blood sugar. If she were to go to a doctor he would run a blood test and find the blood sugar normal because the glucose would be up, but there is a time each day that this person would just feel they are going to die, being anemic, almost blacking out, irritable, fighting for life and wondering why everything is upsetting.

This person's calciums are so deficient that she is a nervous wreck, and except for the grace of God, very hard to live with. Also, there is a potassium deficiency causing her to be extremely absent-minded.

We would find a very high count of delta cells in the vagina, ovary area, and also in the bladder and both kidneys; a very high emphysema count in both lungs, and a case of diarrhea. The food is going through too fast and the person is feeling very weak.

She may have been married many years and become so irritable until the husband can't stand her, and if he has low calciums too, divorce will probably result. He doesn't understand her condition and she doesn't either.

There is a difference between weakness and tiredness, and most people do not distinguish the two. The athletic young man would be unduly tired. Tired in the morning, tired at night, tired all the time, but this woman is weak all the time. This type of person must be in a retreat because unless something is done quickly she is going to be faced with a total hysterectomy because the ovaries are affected. Also, there are delta cells in both breasts. (If the middle finger were placed on the breast it would feel like cords, it wouldn't be like a lump.) Unless something is done this person would have less than two years to live because the energy rating is down to 30. About all a doctor can find wrong here at this stage would be low in potassium, and a very slight calciums deficiency. If he took a pap smear there would be no blood or indication of anything wrong in the uterus area.

The person is probably already over her menstrual period, but the calciums are still too low, giving her the effects of menopause: nervousness, cold hands, cold feet, and in general needing a good tune-up. The liver is not manufacturing vitamin D, and without the liver manufacturing an enzyme of vitamin D the body cannot assimilate calciums. This person must be in a retreat, or place where tests can be done every day. The numbers indicate that there is a tumorous area in the brain. The nerves in the brain are not carrying the messages through, not because they are tired, but because there is not enough electrolyte in the system for the messages to go from the brain to the vital organs. The high count of delta cells in the brain would be in the top of the head, toward the front of the brain, in the medula oblongata. This is the part of the brain with which we think and act, where we store information with which we study and memorize.

There are 4 percent of the people that do not go through a

rough withdrawal, and there's no biological reason. The only reason we know is that our instruments do not measure the grace of God, and, therefore, there is no biological reason why they should not go through just as rough a time as everybody else, but they don't and their numbers change and within four to seven days we can tell whether or not the body responds.

It is impossible for anyone who has not been trained in the Reams Biological Theory of Ionization to understand these numbers. There are seven different stages that we go through from babyhood to senior citizen, and in each stage these numbers indicate something different.

ANY CHANGE IN ONE NUMBER CHANGES THE MEANING OF ALL NUMBERS, THEREFORE, ANY ONE NUMBER BEING PERFECT DOESN'T MEAN ANYTHING, UNLESS ALL NUMBERS ARE PERFECT THEN NO NUMBERS ARE PERFECT.

In this equation stated above there are two ranges that we have not discussed, the B and D range.

You can be just as near perfect health or death in the B range and the D range as in the A range. If anyone's body chemistry was in any range except A, and after a week's dieting their body chemistry moves into range A, this indicates that nature would be restoring you to perfect health at the most rapid rate possible. However, if you started too late and your reserve energy level was too low the disease may be overtaking you faster than you can supply the minerals to repair the damage, this can be determined within a 30-day period.

Range B is the range that indicates that the person really needs a tune-up. The longer this condition exists, the more critical it will become. The reserve energy is the criteria of the exact state of resistance of the body.

Reserve energy between 18 and 60 years of age should be about 100 if one is in perfect health. Many people do not reach reserve energy level 100 in their whole lifetime and may live to be in the seventies, eighties, or even nineties. There are people who break every rule of health and live to a ripe old age. Again, there is no biological reason to explain this, but there is a theological reason. One of the Ten Commandments tells us: HONOR THY FATHER AND MOTHER: THAT THY DAYS MAY BE

LONG UPON THE LAND WHICH THE LORD THY GOD GIVETH THEE. No one knows except the person himself whether or not he or she truly honors (honored) his or her father and mother.

This calculation of energy is very difficult, and only those who master the subject through the ninth seminar will be able to calculate the energy level. There are people now that say they have cracked the code and that they can calculate the energy level. Believe them not. There is no way to crack a code because there is no code. It's simply mathematics used according to the age, height, weight, sex, race, as well as the specimen readings.

In Chapter 7 we said it takes X amount of reserve energy divided by time to make a cell in the body if there is reserve energy behind it. You might ask, "What is the value of X?"

The value of X is determined on the reserve energy by how strong the gastric juices are of one's liver.

I have said many times to the RBTI students, "If you thoroughly understand seminar one, then you can go on to all eight seminars by yourself without help, but unless you do understand seminar one thoroughly and memorize the rules, then you will not understand seminar two, three, four, five, six, seven, eight or nine." I had some of the students in seminar five to say, "Now I know what you meant when you said if we thoroughly understood the math in the body chemistry which was taught in seminar one that we could carry on through all nine seminars by ourselves, but I did not understand it."

We do not mind you seeing your numbers, keeping your numbers, or trying to compare your numbers, but unless you have had the seminars you will not understand them.

It is possible to work so hard until you will still burn up more reserve energy each day than you took in and become more ill with your numbers in range A, and eventually it will be moving into other ranges if you keep this up.

Symptoms of a potassium deficiency in the brain show up long before it shows in the blood, and indicates that the person works like a machine. They never seem to tire. They work long, long hours and very, very hard. Nature seems to tell them that their time is short and they've got to get all their work done and prepare them for death.

A potassium deficiency in the brain means that the brain cannot accept enough oxygen, and if you have this condition and a low sugar condition, you really have a problem because the low sugar lets you know that there is something seriously wrong, but with the potassium deficiency there can be something seriously wrong and you may not be aware of it. Potassium deficiency shows up in range B, C, D, and E. It can show up in any range, but it is most prevalent in range B and D.

Let us take an example in range B:
Female, age 44, weight 120, height 5'5"

| 5.5 | $\frac{6.80}{6.20}$ | 18C | 4M | $\frac{7}{10}$ |

This reading would not indicate a potassium deficiency but would indicate the person being a borderline diabetic, that the pancreas was not producing enough insulin. She would be slightly underweight. The salt number indicates that the cholesterol is just beginning to form and that she's much more tired than she should be, and has a weakness mixed with the tiredness which is upsetting her entire digestive system.

The pH readings $\frac{6.80}{6.20}$ would seem almost correct. However, the liver is slightly too acid, and the body chemistry is slightly too alkaline which would be creating some gas after eating, with minor burping, but not serious.

Because of the borderline diabetic condition, the carbohydrate would rise higher at times and drop back down to 5.5, and if it dropped below 5.5 she would have to have something sweet or the energy would run out completely.

The last numbers indicate the proteins are just high enough that the person would be quite tired all of the time, but also indicate that she is in the menopause with hot and cold flashes, and at times, cold feet. This person is in need of a tune-up.

The numbers indicate that there is a vitamin C deficiency and the longer this condition lasts the greater the deficiency. Also, there is an anemic condition. Most anemic people are much whiter than usual because the liver is not getting enough iodine, iron or calciums.

The reading indicates emphysema. This person is not getting

enough oxygen. One of the finest things to do for emphysema is to walk a mile a day, but don't just start off and walk a mile the first day. Keep on until you walk slowly, one mile a day, with nothing in your hands.

This person's numbers indicate the beginning of collagen disease, or a vitamin C deficiency, which begins to affect all of the vital organs and muscles. Many times there is soreness in the muscles, but it is generally travelling. It's in a joint for a day or two and then another joint and oftentimes for a day in the big toe, indicating that it's gout, which it isn't really, it's just very annoying, and not a very serious problem unless this has continued for a long period of time. The longer this condition exists the more apt it is to cause irregular menstrual cycles, and many times extremely heavy flows, or times in which the cycle lasts much longer. A hysterectomy will not solve the nervous problem and tiredness, but the deficiency can be corrected by correcting the diet.

The reason for the mineral deficiency, the body is not assimilating enough variety of calciums because there is not the resistance between the acids and the alkalines. The body chemistry is out of ratio of one element to another.

All foods that we eat are cationic, with the exception of lemon. This is one reason why we recommend fresh lemon juice water or lemonade: it is nature's form of natural dilute hydrochloric acid, and the liver can take lemon juice when it is taken systematically, and in not too large amounts, and convert it into enzymes with less chemical change than any other natural substance known to man. Then the body begins to take on more calciums and corrects these conditions. The next best substance to use would be a man-made substance: hydrochloric acid tablets. If the mineral deficiency condition has existed a long time, a person will need both the lemonade and the tablets.

Some people are allergic to lemon juice. Those allergic cannot urinate, it stops the kidneys from letting the water pass out into the urinary tract. Also, if one has ulcers it burns the stomach so badly the person cannot stand the burning, then they must be taken off of the lemon and given freshly squeezed cabbage juice three or four times a day (this must be used within 20 minutes from the time it is squeezed), the amounts regulated by the tester, and this will heal any ulcer. I have never seen anyone allergic to cabbage juice that had ulcers. After the ulcers are under control, six weeks to three months, then the person can start on the lemonade and it will rebuild the liver and almost make them look like they have had a bath in the fountain of youth.

The D range is a more serious range than the B range because the body is extremely too acid, causing extreme nervousness, abnormal fear, and except for the grace of God, almost a maniac. One would have the same problems as in the other ranges but more exaggerated than in the A and B range.

An example in the D range could read:

$$0.9 \qquad \frac{5.80}{5.20} \qquad 12C \qquad 4M \qquad \frac{7}{5}$$

This condition is one that extremely upsets the nerves. The numbers indicate a potassium deficiency. The message is not going through from the brain to the vital organs, and it's almost impossible to work. Such a person has an idea that everyone is against them—there's a chip on their shoulder, and they are sick, and yet they don't look sick. They can also be anemic, and if this were a woman, from puberty to age 60, could be very deep in menopause with this number combination. The suffering is extreme because the nerves exaggerate the condition. THE LOWER THE CALCIUMS, THE GREATER THE FEAR.

The numbers show:
- advanced delta cells in the vagina, uterus, ovary area, and in both breasts
- a high emphysema in both lungs
- a cystitic condition
- the kidneys are not functioning normally because the liver is not manufacturing enough enzymes to maintain the energy needed to restore the energy used each day.
- have to drive themselves to do everything that they do. It would be unsafe for this person to drive a car because the blood sugar is so low, and this can cause as many accidents as people who are intoxicated by alcohol while driving. It slows down their emotions. They may temporarily black out, and don't know it until it's too late.

ONE OF THE REQUIREMENTS IN OUR DRIVER'S

LICENSE TESTS SHOULD INCLUDE WHETHER OR NOT THE PERSON HAS LOW BLOOD SUGAR.

Urinary tests are more accurate than the blood test on low blood sugar because the urinary carbohydrate reading equals the blood chemistry readings over any 24-hour period.

These RBTI test numbers are not easily interpreted. This book is written to educate the public to the place that they will go to someone who has trained at the seminars in the Reams Biological Theory of Ionization.

Some of the testers are so steeped in fads until it takes two or three years to de-fad them, but don't hold that against them. I would rather trust my life to the poorest tester in the field than to trust my life to the use of chemotherapy, cobalt or radiation.

Chapter 18

RBTI AND PASTORAL COUNSELING

Many calls have come from pastors begging, "Please do something to help me. I do not know enough about the health message to teach my people how to be healthy. I have just lost with a heart attack one of the finest members of the church, and a very heavy financial supporter, in his thirties. He is a great loss to the church and to God's work."

A Baptist pastor in Florida said, "Please, do something to help us. I have just lost one of the fine supporters of our church, only 45 years old with cancer. He had cancer for two years and the cobalt seemed to hasten his death. It breaks my heart to see that. When he was told he had cancer he asked me as his pastor what I thought about it, and I agreed that he should do what the doctors said. Now since we have been enlightened on the dangers of chemotherapy, radiation and cobalt and know it has hastened his death, his wife has left the church. She pointed her finger at my nose and said, 'If you would have counselled us correctly, my husband would be living today.' Reams, you've got to help us."

We are establishing in the very near future seminars to teach pastors how to counsel their congregations on health. We are not going to tell them not to have cobalt, radiation or chemotherapy, but we are going to teach them God's way to being healthy. The use of drugs is something for the doctor to convince the people, and not the pastor. Ministers need to be taught in the seminaries the Bible health message.

Pastors are demanding us to feed them the health message in the Bible. We are not going to give them the same teaching we teach the doctors, chiropractors and other testers and counsellors. We will teach them to counsel their people, of all ages, in regard to their health, as given in the Word of God.

Many ministers today are deeply concerned because of the many divorces among the young, and even senior citizens. Many times divorces are brought about because the calciums are too low in the system and the individuals are a nervous wreck. Husbands and wives are at each other's throat from morning to night. They love each other and each one is fighting for his or her life, but they have no one to tell them what to do to correct these calciums deficiencies. Ministers are not to tell the people which calciums to use but they will be informed on doctors trained in this field to which they can send their parishioners for help. And this will prevent many, many divorces.

In 1972 at 11 o'clock one Sunday night my doorbell rang. It was the pastor of our church in a small city in Florida. "My wife is having a heart attack and is almost unconscious. Please help me quickly," he begged. I got my bathrobe and bedroom slippers on and went with him to the car.

His wife was a rather large woman. There was no way at this time that any tests could be made, but I quickly examined her hands and feet and found that they were cold, and by this I determined that it was not a heart attack, however, she was having some severe chest pains, the heart had lost its rhythm, and had dropped down into the lower 40's. The nearest hospital was 15 miles away and I was afraid that she would not last to get to the hospital, so I made up an ordinary drink for the purpose of relieving gas on the stomach and told her to sip it slowly for 30 minutes. Within 10 minutes the gas began to burp out of her system. In a half hour she had perfect relief, the heart rhythm was back to normal, her feet and hands warmed up, and she felt as if she had never been sick. I asked them to come to the lab and I tested her to see what caused the attack of acute indigestion. Her calciums were very, very low, which is one of the causes of acute indigestion.

I said to the husband, "You need a test also because there may be a problem here that involves both of you." She was in her mid-thirties, and in menopause, and he was 38. His test indicated very low calciums also. The reason I thought he would have low calciums was that the wife did the cooking and if she was low in calciums then probably he was too. This is what we can call deduction in mathematics. If you find one problem in one member of the family, then you should look for it in other members of the family, especially if you find it in the mother and she's the one who does the cooking. There are exceptions to this rule. Nevertheless, it is absolutely necessary to go by the numbers. Do not try to draw any conclusions without scientific facts.

After I found him low in calciums I said, "Most people that have your numbers would have a very difficult time living with a wife because you would be at each others' throats every minute," and both of them looked as if lightning had struck them. Then they admitted they were on the verge of a divorce because they couldn't get along with each other. They had three children. I said, "It is very easy to fix this so that you can get along with each other." I had the calciums they needed and gave it to them to take, and in two weeks the nervousness had completely left them and they were deeply in love with each other again. Now whenever either one gets the least bit cross the other runs for the calcium bottle and says, "Here, honey, take your calciums!" I also told the wife what to do so she would not have acute indigestion.

So you can see how important it is for people to have a dietary test made that will correct the mineral deficiencies in the body.

A young married couple can have the same problems with low calciums and their life becomes upset and it leads to divorce because the low calciums had so upset the wife's nerves she cannot forgive and forget as quickly as a man, and neither can the man when his calciums are too low. When calciums are normal and all of the numbers substantiate that they are normal, it is very easy to forgive. The lower the calciums, the more difficult it is to forgive and forget.

During the act of intercourse there is an electrical charge that passes between the male and female. Many times when the young couples get married, it takes up to five or six months to learn the real art of intercourse to get the maximum amount of ecstasy that is normally due each other. There are conditions wherein the calciums are too low in one or the other and the sex act is a trying experience, and yet the grace of God keeps them from being irritable.

We now come to the problem that the Bible speaks of as fornication.

Those who engage in sex without marriage have problems. There is a guilt feeling because they are not married, and they have not had enough practice in the act in order to reach the heavenly harmony that was intended by God for the man and his wife. Sex acts without love is very bad because it leaves each torn down rather than built up. It causes the sex organs to refuse to accept enough manganese to keep the prostate in a man from having too many delta cells. It also causes a larger count of delta cells in the vagina area and sometimes the breasts of women, and it upsets the entire nervous system because it tears down the calcium contents of the system rather than build up the reserve energy, and decreases the reserve energy greatly.

God made men to have one wife, the wife to have one husband, and any breaking of this law is an attempt to break the biological laws made by the Lord Himself. In reality no one can break these laws, they only break themselves on these laws. Consequently, those women that sell themselves to prostitution for money, or any other reason, simply look old at 30. They have burned themselves out for a few unknown reasons, sometimes money. These acts cannot be stopped by legislation. They can only be forced on the black market conditions, and it is up to the ministers to teach the young people the dangers of fornication, and not for the state to try to punish them for their acts.

If more young people were taught how much damage is done by fornication they would listen and obey the laws because they want to be healthy. I'm not saying that fornication is to be avoided in all circumstances from a biological viewpoint. There are women and men that are said to be over-sexed, and for some reason they do not marry, or their calciums are too low, and consequently they cannot find a partner that suits them enough to be married and live an honorable Christian life. The passion drive is so great in some that it upsets the calcium balance and the more the calcium balance is upset the weaker they become, and the weaker they become, the less faith they have, and the less faith they have, the more undesirable they become as a prospective partner. Many women are in an insane asylum because of this very condition of sinning. They do not even understand their problem.

Here again if a pastor knew how to counsel (biological church problems) such people instead of telling them they are going straight to hell and would teach them to go to someone who can build up their calciums, probably have them take a Dale Carnegie Course to improve their personality, and to have a class in which the facts of life were presented to them in a Christian way, that many converts could be won to God's Kingdom.

This does not mean that the pastor or the church is condoning the illicit use of sex. The Bible says there is none perfect, no not one. It also says that he who does not keep the commandments and teaches man to do so is least in the kingdom of heaven. But he who keeps the commandments and teaches others to keep them is called great in the kingdom of heaven. *The church is a hospital for sinners and not a clubhouse for saints.*

Jesus asked the woman when she was accused of adultery as he wrote in the sand, "Where are thine accusers?" Then said to her, "Go, and sin no more." Jesus did not condemn this woman, neither do I condemn her, but I put these people on the top of my prayer list. Anyone that would put such a person down, does not deserve to be in the kingdom of heaven. It is our sole duty as Christians to labor until these souls are born into the Kingdom. Every soul that is born into the Kingdom is there because of somebody's prayers and never because of somebody's gossip.

Please do not say, or quote me as saying, that I condone fornication or adultery. I do not condone it. However, it does exist, so there must be a way that God has to solve these problems, and I believe that much of it can be solved with God's love, our love and biological means. In another chapter you were informed that women use seven times more calciums every day than a normal man, and for every time that the husband goes home and his wife is in a very bad humor and punishes him by refusing him the love, the bed life that he is entitled to, that is when he is going to go to someone else and he's going to say, "My wife doesn't understand." (How fortunate he is that his wife doesn't understand or she would throw him over the sandy bank.) Now this does not justify him whatsoever, but it is a pastor's duty to know how to counsel and to take care of his sick sheep. And these people are sick and any shepherd that does not know how to care for his sick sheep is not going to have them very long.

The pastors' seminars are to teach the pastor how to feed his sheep and how to care for his well and sick sheep so that souls can be won to Christ, and so that the young saints can grow up to be strong, mature saints of God.

Chapter 19

OXYGEN

Creation begins with the Garden of Eden and ends with a Garden. As long as anyone tries to bring God down to his size, he will not understand the beauty of creation.

Many people think that Adam and Eve were naked in the Garden of Eden. They were not naked until the fall. They were clothed in light we are told in the Book of Psalms. One of the interesting things about Adam and Eve as they were clothed in the light, was that they were in the image of God, and the Word says God is "a flaming fire, a flaming torch," and in the Book of Revelation it says when we are translated we shall be "like Him, we shall shine as the sun."

One thing about man being created in God's image, we were created without any blood in our veins, or we were created without the need to breathe oxygen. If we had breathed oxygen we would have exploded and would have caught fire. According to the mathematics of the frequency upon which we live, the temperature of Adam and Eve's body in the Garden of Eden before the fall was 980 degrees centigrade, because their bodies were anionic.

Adam needed a helpmate so God gave him Eve.

The word ADAM means "red," for Adam was made from the dust of the earth, and it was red. Eve was taken from the rib of Adam, and male and female created He them.

Some believe that the tree in the midst of the Garden was the tree of sense. I believe it was a fruit. As Eve partook of this fruit she was tempted by the king of all beasts, the dinosaur (serpent) and he was commanded to crawl on his belly because he deceived Eve. The dinosaur became the snake after he lost his legs and was not in the Ark, so we are led to believe that the flood completed the extinction of the dinosaur.

Adam had a choice of either eating the fruit and having a wife, or he had a choice of not eating the fruit and losing his wife. He chose to partake of her sin and have a wife. Adam's sin was: he listened to his wife, instead of listening to God. God made man the head of the house, the prince, the high priest of the home.

I've never seen a happy home when the father was not the high priest of that home. It is his duty to lead his family to the Lord. That's why Abraham was such a great man, for he led his family in the ways of the Lord.

After Adam partook of the fruit he realized that they were naked, and a change had taken place in their body chemistry. It changed from anionic to cationic. There was now blood in their veins and there was sweat, they would earn their bread by the sweat of their brow, and there was death and they began to die.

Any day that one breathes oxygen that day they begin to die. Oxygen is known as the element of time. It is the element that brought time into existence because where there is no time, there is no oxygen or death. You might ask: "If there was day and night is that not time?" No. *You can have rhythm without time, but you cannot have time without rhythm.* Adam had lost his anionic body and was living in the age of time, which denotes an end. Time denotes a death. Have you ever had an appointment and somebody didn't show up? This created a problem. Time denotes problems. So here then is another proof that Adam and Eve breathed oxygen because oxygen is known as the sword of all elements. Everything that comes in contact with oxygen begins to break down. Metals begin to rust and deteriorate. Apples cut open begin to turn brown.

Man cannot live without oxygen. What did Adam and Eve breathe before they breathed oxygen? The Word tells us when God breathed into man the breath of life he became a living soul, and this was while he was still in the Garden. God breathed into Adam the Spirit of God, the Spirit of Love, the Spirit of Divine Supremacy of something that is greater than we can express in any human language. Adam's dominion never used oxygen, neither did the fish of the sea or the animals, nor did the trees need carbon dioxide. They lived on the atmosphere of God. So when Adam fell his whole dominion fell. Before that trees never died and never shed their leaves.

Adam was the largest farmer that ever farmed the earth, that is, before the fall. His farm was approximately five times the size of the United States and in Adam's anionic body he was not limited by time or space. He could go great distances instantaneously. This is hard for us to conceive of now. Suppose you wanted to make a telephone call from where you live to the farthest opposite city on the North American Continent. You could go there and tend to the business and come back quicker than you could get a telephone call through. This gives an idea of how fast the anionic body can move. An anionic body could be all over the Garden of Eden each day.

When Adam and Eve began to breathe oxygen their body chemistry changed and they had cations, which meant a loss of mineral, which meant a loss of energy. When the earth was young there were minerals everywhere and the people lived to an old age. Adam lived 920 years and Methuselah lived 969 years.

When the three Hebrew children, Shadrach, Meshach, and Abednego refused to bow to the golden image set up by King Nebuchadnezzar, they were bound and cast into a burning, fiery furnace. It had been heated "seven times more than it was wont to be heated." The flame of the fire slew the men who placed them in the furnace, yet when the king looked, he said, "Lo, I see four men loose, walking in the midst of the fire, and they have no hurt; and the form of the fourth is like the Son of God." When they came out their clothes were not burned, or even smelled of smoke.

When Elijah was taken up to heaven in a flaming chariot—a chariot of fire—bears out what I'm saying about the temperature of God, the Father and the Son.

When Jesus was crucified a sword was thrust in His side and His blood and water ran out. He shed the last drop of blood He had for your sins and my sins and then was buried, and on the third day He arose again. He went down in a cationic body and He arose in an anionic body like Adam and Eve had in the Garden of Eden before the fall.

Let's compare the cationic body we have and the anionic

body our first parents had. Whenever we eat food it has to be mixed with the saliva in our mouth, then it goes to the stomach where it is digested by the gastric juices manufactured by the liver and the pancreas, and we live off of the energy of the food we eat, and the excess of this food is passed out of our system, excreted from our body, and there is decay. This food that passes out of the body is in the state of decay.

Leaves fall from the trees and decay. When animals die they decay. This is because we are living in the oxygenic age. But in the Garden of Eden, before the fall, Adam and Eve ate fruit, but the fruit was digested by the heat of their own bodies and the energy, or heat and electrical energy, went right out of their bodies. They had absorbed the total amount of energy that was in their food and therefore there was no decay. Like putting wood into a stove and the stove giving off heat, and yet there was no smoke. There were no bowel movements. They didn't have to urinate. There were no tears. There was no death. They were living in the age of eternity, in which there was no time.

Jesus, after He arose from the dead met with some of the disciples and ate fish. That fish He ate went off as heat energy and electrical energy because He had no blood. There is no constant frequency in the blood, it is only our cationic flesh that has frequency. I Corinthians 15:50 tells us that flesh and blood shall not inherit the Kingdom of God. Therefore, it cannot enter the Kingdom of Heaven. So we will have a flesh but not the cationic flesh. We will have an anionic flesh, and we too will again be clothed in light which will appear as a white raiment.

There is another important factor that bears out this biological science of the resurrection given in the Bible. Hell is fire and brimstone. Any high school student who has studied chemistry knows that brimstone is sulphur, and that sulphur burns at 720 degrees centigrade, and that sulphur casts off heat fumes which displaces oxygen, which would make it impossible for anything that breathes oxygen to live.

Let us draw a conclusion now that heaven is hotter than hell, but the saints can walk through hell and not even need a fan as their body temperature is 980 degrees, and hell only has a temperature of 720 degrees. So we will then have become like the Father again, as Adam and Eve were in the Garden of Eden.

Satan tries to make everyone fear death. He tells big lies about death. He tells you that when you are dead you are going to be dead a long time. That is not so. In death there is no time. The person that died in Abraham's day, or before, will not have been dead any longer than the person who dies one hour before Jesus comes, because out there, there is no time. The Book of Revelation tells us, "time shall be no more." When you go to sleep at night and wake up the next morning, it's as though no time has passed. The event that revealed this so clearly to me was during the war when I was unconscious for 31 days and then I awakened. The last memory I had was that I was thrown above the trees and my last thought was "This is it." I was sure I was going to die, and the next thing I knew was 31 days later, 2,400 miles away in the Seventh General Hospital in New Guinea on an operating table, and it was just like seconds because I said, "I sure did land easy." The 31 days to me was only a matter of a part of a second. I knew nothing of what happened during those 31 days. So in death there is no time.

To be absent in body is to be present with God, or, to await the judgment day of the soul. We don't exactly understand it, but we do know that each of us has our own number as frequency, micronage, milli-micronage, or milli-milli-micronage, as any chemical has in chemical analytical numbers. When Jesus comes and we use the password that we've been told in Isaiah 25:9 to use when we see Him, "This is our God, we have waited for him," we shall go down in corruption, but shall rise in incorruption—then we shall be like Him, we shall be changed in the twinkling of an eye.

It can be assumed from the following verses that we will stand before God and our name, or "numbers," will be called:

Daniel 7:10: "... the judgment was set, and the books were opened."

Daniel 12:1-2: "And at that time shall Michael stand up, the great prince which standeth for the children of thy people: and there shall be a time of trouble, such as never was since there was a nation even to that same time: and at that time thy people shall be delivered, every one that shall be found written in the book. And many of them that sleep in the dust of the earth shall awake, some to everlasting life, and some to shame and everlasting contempt."

Revelation 2:17: "... To him that overcometh will I give to eat of the hidden manna, and will give him a white stone and in the stone a new name written, which no man knoweth saving he that receiveth it."

Revelation 20:12 and 15: "And I saw the dead, small and great, stand before God; and the books were opened: and another book was opened, which is the book of life: and the dead were judged out of those things which were written in the books, according to their works. . . . And whosoever was not found written in the book of life was cast into the lake of fire."

Revelation 21:27: "And there shall in no wise enter into it any thing that defileth, neither whatsoever worketh abomination, or maketh a lie: but they which are written in the Lamb's book of life."

There are times when I personally have wished I had nothing to do with the healing arts. I have had to watch my prayers go unanswered. I have had to watch people as they departed from this earth to the world beyond, and it is not a pleasant thing. It is the doctor's business to console the parents or the loved ones. I've always tried to do this when I knew by their numbers even days before the person passed into eternity that we were losing the case. I was there to tell the loved ones to think of the passing of this person as them making a trip to the Holy Land, like going to Jerusalem, or to the Sea of Galilee, and coming back and telling you all about it, telling you where Jesus walked and what He did. Only this time they are going to the New Jerusalem where Jesus really is instead of where He was, and they are not coming back. Some day you will meet them and they will be waiting, ready to greet you. When you look upon death as being a promotion, promoted to glory, it is very beautiful.

You might say, "That's very easy for you to tell someone else, but when it actually comes to your own it will be a different thing."

I had a son killed by a drunken driver and this is the way I took it: God needed him in His Garden and He took him and I'm looking forward to the day in which I shall see my son. Also, when my mother and father passed away I took it the same way.

It is the easiest thing in the world to die. It is just like going to sleep. No one really knows when they die, but they will awaken and be looking in the face of Jesus, face-to-face! I have no fear of death. People that weep when they lose a loved one are selfish. They think they know more about life than God, and many times they become bitter and hate God. God is seeking to draw them a little closer to Him. Who am I to question the ways of God?

I have tried to rejoice with those who rejoice, and weep with those who weep. I have tried to live in a house by the side of the road and be a friend to man.

Chapter 20

DIGESTING FOOD

If you do not love your country enough to die for it that your children might have liberty throughout their lives, you do not love your country. If you do not love your children and your grandchildren enough to die for them, you do not love them, neither do you love yourself. If you don't love yourself, you can't love God. This is what it means to be an adult saint.

We are to obey the laws of the land as far as it is within us without disobeying the laws of God. But if there is a conflict between the laws of God and the laws of man, then we are to obey God. This was the very reason that Paul was beheaded. He was told if he would renounce Christ, the Messiah, as his Lord and King, that his life would be spared. He refused and he was beheaded.

The kingdom of God is within you and you can't love anyone more than you love the God within you. Before you can hate anyone, you must hate yourself. With the same hate that you hate yourself, you hate others. **Hate is the finest cancer seed in the world.** No wonder that 25 percent of the American people will die of cancer.

Do you know that it costs $40,000 to die from cancer with most medical doctors, and that you can die for free with cancer without radiation, without any cobalt, without any chemotherapy, and without any false hopes, and with a lot less suffering? Ask anyone who has taken the chemotherapy treatment.

Do you know that it is possible to live to a ripe old age by obeying the laws of God? The health message begins in the first chapter of Genesis and goes throughout the Bible. In First Timothy 4:1-5 we read: "Now the spirit speaketh expressly, that in the latter times some shall depart from the faith, giving heed to seducing spirits, and doctrines of devils; speaking lies and hypocrisy; having their conscience seared with a hot iron; forbidding to marry, and commanding to abstain from meats, which God has created to be received with thanksgiving of them which believe and know the truth." (If you know the truth then you will know the eleventh chapter of Leviticus where we are told which are the clean and unclean meats). Paul, a Jew, was writing to Timothy who would not think of eating any of the unclean meats, and when he says "every creature of God is good, and nothing to be refused, if it be received with thanksgiving," he was talking about the clean meats that were offered to idols. There was quite a discussion among Christians in that day whether or not they should eat things that were offered to idols, or whether they should eat the sacrificial meat that was given to them by the priests. This was not referring to the hog, shrimp, oysters, lobster, catfish and etc., because in the verse before it says, "which God hath created to be received with thanksgiving." God did not create a hog, or any other unclean meat, to be eaten with thanksgiving. If God is the same yesterday, today and forever, then God doesn't change either in the Old or the New Testament. If God changed what it said in the Old Testament to what is said in the New, then God is a God of conflict and doesn't know what He is doing.

"For it is sanctified by the word of God and prayer" (v. 5), this speaks of the meat that was offered as a sacrificial offering. It was sanctified then by prayer, but there is no way to sanctify hog meat by prayer when either Jew or Gentile eats it, for pork digests in three hours and burns up in your system too quickly, bringing old age on too rapidly, bringing you to an early death. This is true of all the unclean meats. So there's a scientific reason that proves the explanation of these words to be accurate.

Where the Lord spoke to Peter about the "unclean meats" and told him to eat thereof, Peter said, "But, Lord, am I to eat the unclean meats?" And God said for him to eat. God was telling Peter to take the message of salvation to the Gentiles. The Gentiles were looked upon then by the Jews as the black people were by some Americans 40 years ago. Jews felt that they were so much better than the Gentiles and that the message of salvation was for the Jew only. This vision was to tell Peter to take the message of salvation to the Gentiles, and does not refer to food.

If you take Scripture out of context you can prove anything you want to prove. But there is no way that you can eat the unclean meats without reaping a premature grave, and also have a rather sickly journey in so doing. A sick gentile in the New Testament is just as sick as a Jew in the Old Testament.

In Isaiah 66:17 we read: "They that sanctify themselves, and purify themselves in the gardens behind one tree in the midst, eating swine's flesh, and the abomination, and the mouse, shall be consumed together saith the Lord." "For they that sanctify themselves," is like saying: "Many people do it and if they do it, I can too." In other words, they justify themselves.

This also speaks of our leaders, they are spoken of as being a tree. In Psalm One we are told "He shall be like a tree planted by the rivers of water that bringeth forth his fruit in his season; and whatsoever he doeth shall prosper." Psalm one is speaking of a man who is happy because he has not listened to the counsel of the ungodly, has not sat in the seat of the scornful, and that he has found favor with God. He is classified as a tree.

A weak person says, "Because this great man eats pork it's all right for me." Maybe he's the preacher of the church, a Sunday school teacher, or a Sunday school superintendent. Holiness means absolute obedience, and sanctification means that I will walk in the steps of Jesus regardless of what it means to me. I will walk a little closer to Him today than I did yesterday so that I may be my brother's keeper.

I love my country and I'll die for it if necessary to preserve the liberty of America. I will not be threatened by the devil worshippers. I will do everything I can to keep the devil worshippers from taking my children. I will not be threatened by the homosexuals in all of their organizations demanding civil rights. I will do everything I can to keep homosexuals from being school teachers. I will do everything I can to protect the very foundations of the Constitution of this country.

Unless you believe that a religious nation and the Bible are inseparable, you are being deceived. The devil says separate the church from government. If you separate the church from the government then you can control the government and the government can control you and close all the churches. It is impossible to close the churches, separate the principles of the royal law from the government and retain our liberties.

Let us now stand firm in the faith.

Let us now stand firm for liberty.

Stand firm for a referendum of all our laws, and obey the rules, regulations and suggestions given in this chapter.

This chapter does not deal directly with the Reams Biological Theory of Ionization. However, it has everything to do with your digestive system because if you will stand each day for the rights and liberty of yours and others, then your food will digest better. You will become stronger because of the problems that God has given you, and you have laid them at His feet and He has handled them for you, or you will become weaker from running from the problems, and the more you run from the problems the more you will worry about the problem. Worry is found cancerous in all souls.

The Bible says "Because strait is the gate, and narrow is the way, which leadeth unto life, and few there be that find it" (Matt. 7:14). Are you willing to be among the few, or are you going to be among the masses of lukewarm Christians that will be spewed out of His mouth? When the Lord comes, are you going to be the one who says, "Lord, did I not do this for You?" He is going to say, "Depart from me you cursed, I never knew you." Unless you are ready now to stand up and be counted as a follower of God and proclaim Him before all men, He will not proclaim you before His Heavenly Father (Luke 12:9).

You might ask, "Why do you continue in this field of the healing arts in the field of diet? Why do you care whether America lives or dies? At your age, does it really matter to you? What is your motive in defying those people that you say falsely accuse you? What do you hope to gain?"

Jesus has commanded us to be fishers of men, not only for men's souls, but for their faith, for their works and for better health. He also said to "show thyself approved unto God, a workman that needeth not to be ashamed, rightly dividing the word of truth." This I do, to the best of my knowledge. Also, Jesus said, to seek to know the truth and "the truth would make you free." Then, too, He said in Luke 10, and in other places in the Bible, "to heal the sick." So this is why I do what I do. This is why I enjoy the persecution, because I am obeying my Lord and Master. I will serve Him as long as I have breath. If I have liberty I will serve Him wherever I am. If I am confined to a prison then I will start a revival meeting and preach to the lost and serve them with whatever I have. Even though I be persecuted, I will not hate. For I know that those who persecute me do not know what they are doing. There is not enough gold in Fort Knox for me to trade shoes with them. The Bible says, "if they persecute you in one place, flee to another."

Jesus did not have a place to lay His head, neither do I, except as friends offer me a place here and there.

I will do everything in my ability to make America the healthiest nation on earth, to keep our children from being sick, and from having unnecessary operations. There is a time for operations, but the operations should be a means of last resort.

- Do you not know that all illnesses are caused by a mineral deficiency? The cells simply begin to decay because there are not enough minerals to replace those cells that are worn-out.

- Do you not know that heart attacks are caused because the worn-out cells are not washed out of the system and the salt within the system is turning it to urea, which causes the heart to beat too hard and sometimes causes pain in the left arm, in the chest, and then ends in a pectoris heart attack? (The word PECTORIS is a Latin word which means pain in the chest. However, I am classifying the word pectoris to mean only a heart attack because of too much urea in the system.)

- When our bodies retain too much salt, the blood vessels lose their ability to expand and contract and cholesterol forms in them, and it causes the intestines to lose their ability to expand and contract and they form pockets or diverticulitis. Nature is robbing from one part of the intestines to supply mineral for another part.

- This is also the cause of cancer of the bone. There is not enough mineral to supply the other elements so nature will rob a bone here and there in order to supply the mineral to maintain life.

- Women lose their teeth in pregnancy because there is not enough mineral to supply the bone of the fetus, so nature takes from the mother in mineral and puts it in the fetus; consequently, there are cavities.

- Bones break down because the mineral in our foods are deficient today.
- A number of diseases today are attributed to lack of minerals and lack of exercise. We cannot take the roughness and exercise that our forefathers and fathers took. We ride everywhere we go, and consequently we do not take enough exercise. Hysterectomies could probably be omitted if women did more walking. Their menopause conditions could possibly be prevented in a large measure if there was more walking.
- Cancer of the prostate could likely almost be eliminated if people did more walking. By walking we get more minerals from our foods. The more we walk, the more energy we use. If we do more walking than we take in minerals, then walking could do us harm.
- Many of the above mentioned items have to be under supervision and under the guidance of instruments that are available to us today. IT IS EASY TO BE WELL!

At the present the RBTI test has not been approved by the drug establishment. However, there are no laws that say that anything has to be approved by anyone. If there is such a law then we have lost our liberty. However, there are dangerous drugs, insecticides, and chemicals that should be even more closely guarded than is being done.

We cannot separate life from religion, nor can we separate health from religion. The Word of God tells us: bear ye one another's burden, and that we are our brothers' keeper.

Surely religious people are happy people. One of the problems with Americans today is its citizens are not happy. There are a number of reasons:

They have lost faith in the church as being the agent of healing.

They have lost faith in those medical doctors who give cobalt, chemotherapy, radiation and drugs and surgery unnecessarily.

They have lost faith in our government because churches are permitted to be formed under the guise of religion, such as devil worshippers. Devil worshippers are deceivers to this country and to God. The Bible tells us that "the old serpent is a liar and the truth is not in him." So any church that is devoted to devil worship, you cannot depend on what they say.

It is sad that the government permits sabotage in the name of the church and yet persecutes the saints who are trying to make America the healthiest nation on earth. Does this make sense? Is it logical? Is this constitutional? Do we not still have the freedom of speech? Do we not still have the freedom of religion to worship as we see fit? Do we not have a right to teach the health message as it is written in the Bible? It is the drug establishment that says, "If you do anything to make America more healthy without drugs, I will put you in jail." (These poor doctors who have been so brainwashed to use drugs will not be held guiltless when they stand before God.)

An officer of the law said to me sneeringly, "We cannot keep you from worshipping the Lord as you see fit, but we can harass you 'til you wish you never had." In harassing me they really do not harass me at all. They only harass themselves. You cannot break the laws of the Bible. You cannot break God's laws of physics, chemistry or mathematics. You can only break yourself on those laws.

I have seen the wrath of God come upon approximately 50 people. God blotted them out like you and I would a cockroach. When the adult saints are being persecuted, they do not hate the persecutor, they feel very sorry for him. They know exactly how to get even with their persecutors. They put their name on the top of their prayer list and there is no way for that person to get off. They pray for them without ceasing. They pray for them that God might do for them what He did for Saul on the road to Damascus. That He might reveal Himself to them so that the person that was once their enemy could be their friend and could go into eternity and live forever as a friend.

The devil numbers his children by hate, but God numbers His children by love. God the Father is love. The Holy Spirit is love. The real saints of God are love. True love is a life of service. It is a life of enduring. It is a life of being a servant whether appreciated or not.

I am now a senior citizen. My children are all grown. They are all self-supporting. However, there are times when they do need guidance, or suggestions for guidance, and they ask for it. But they could very well get along without me.

I am now financially independent enough that I do not have to do this work for my daily bread. I served 38 years in agricultural engineering and it was through that that God was good enough to make it possible for my retirement in 1968. I wanted to go into research to know more about the mysteries about the laws that God has made.

God began to send to me for help the ill, the cripple, the lame, the dying, that medicine had failed to cure. The Bible says, If your neighbor asks for help, give it to him and as often as "you do it unto the least of one of these my brethren, you do it unto Me."

I have made many errors and many mistakes, but they are mistakes of the mind and not mistakes of the heart. God has turned these mistakes into blessings.

I am now teaching this Reams Biological Theory of Ionization seminars to qualified people, in order to help carry out this vision of making America the healthiest nation on earth. . . knowing that the history of American liberty is written in blood!

Chapter 21

THE GREATEST LOVE OF MY LIFE

You must be born again.

Today I died. I cried until I laughed and laughed until I cried and cried until I laughed, with joy! . . . because I had left my sardine-can world for a universe. I had left my sinful ways and God had permitted me to be born again into His Kingdom. This was no laughing matter at first; I was sorry for my sins. My heart was broken, I was filled with grief, and then the Holy Spirit came in and took away all the grief! I don't know what He did with it, but it's gone. It's gone . . . and now I am free!

God is truth. The Scripture says, Seek to know God, and "He shall make you free," and this is what I was weeping about, because a new life had begun. I was only an infant, yet an adult in years, but a babe in Christ. No wonder I cried . . . because of the joys I had missed, of the liberty and the freedom that I had spurned. The things that I had once loved, I had no more desire for. The things that I once had no desire for, became as a beautiful garden of flowers. The ecstasy of the Spirit of God within me was like looking into the face of a lily or an orchid, or being lifted on the wings of a hummingbird. I was truly partaking of, I was suckling on the manna from Heaven, and drinking from a fountain that would never run dry.

The book I had so often read as a history book became a message from home. It became a guide, it became a map. It showed me where I was in life. It showed me where I wanted to go, and I then used that map. I still use it and it is like honey in the honeycomb.

It is nourishing.

It is surprisingly lasting in its joy.

It is a history book.
It is a physics book.
It is a mathematics book.
It is a religious book.
It is a book of law.
It is a book of domestic tranquility.
It is a book of health.
It is a cook book.
It is a book of fiction.
It is a book of truth, divine truths.
It is a book of heroism.

It is a book for youth, for teenagers, for young married couples and business men and business women.

It is a book for executives, business managers.

It is a book for school teachers, telling how to teach and what to teach.

It is a book of disappointment, of grief, of sins of the people.

It is a book about farming, gardening, transportation, seamanship, vessels, boats, trees, forestry, fish, birds, beetles and bugs.

It is also a book of geography.

It is a book about astronomy, about the creation, the formation.

It is a book that no man has ever yet understood all about.

It is a book about construction, about building of temples.

It is a book about psychology and how to win friends and influence people.

It also tells us how to do architecture, and why the Tower of Babel fell (because of the poor architecture that was in it).

It tells about what would happen when we disobey the law.

It tells us about the rewards after this life.

It tells us about God.

It tells us about His Son, Jesus Christ.

It is the first and only book that ever told us anything about Him. The rest are just counterfeits or duplicates of it.

It tells us prophecy and predicts the future.

It is clairvoyant in the fact that the clairvoyancy is the supremacy of God.

It is the finest book on astrology that was ever written.

Instead of asking the stars what the future holds, it tells us how to go to the Father himself and get the information firsthand instead of secondhand.

It is a book of stories, mistakes, heartbreaks, rejoicings.

It is the story of humanities.

The Bible, the Book of ALL books, the only *lasting* book, tells about man's ways and how they differ from God's ways, and how God's ways can be in man and how man's ways can be God's ways.

It is a book that I highly recommend to be read.

It is a book that tells us about wisdom.

It tells about knowledge.

It teaches anyone how to be rich, infinitely rich.

It teaches anyone how to be obedient.

It teaches how to be a servant, how to be a workman that needs not to be ashamed, rightly dividing the word of truth.

It teaches us how to defend ourselves.

It tells us what to do in time of trouble.

It tells us what to do in time of peace.

It tells us what to do in time of war.

It tells us how to prepare for the great tribulation.

It tells us about the destruction of the great cities in a day of the atomic bomb.

It tells us about a King who is going to rule this whole earth. Many men have tried to become the rulers of the whole earth; all have failed! but this new King who is coming, He is going to be the King of this planet as well as the universe. There are souls that are so small, that are limited to this universe and to this planet, but there are going to be souls that are not limited to this planet or this galaxy. They are going to be able to be sent on missions, and shall travel faster than the speed of electricity or the speed of light.

This book also tells us of, or predicts the use of drugs. It predicts the fall of churches. It predicts that some men will be lost.

It is the greatest book I have ever read. I don't *use* it; I memorize it! I carry it with me and I mark it. I leave something to show that this passage has blessed me, and I hope it blesses you.

I recommend this book because it has given me life anew, and it did something that no other book has ever done. It showed me that I did not have to die to go to Heaven, that I am there already, but one day I shall be promoted to glory. This cationic body will be transformed into an anionic body. This raiment that I have on will be burned to a crisp, and I will be clothed in light. I will be clean, perfect again. I will be made young again, and I will never grow old. This book called the Bible did all of this for me. If it did it for me, and I am nothing, what can it do for you?

The Bible has taught me how to cleanse the temple.

It taught me how to be fearless.

It taught me how to be filled with holy boldness.

It taught me how to drink pure distilled water from the fountain that shall never run dry.

It taught me how to sing, even though I've never learned to carry a tune. It did teach me how to make a joyful noise unto the Lord.

It taught me how to love the unlovable.

It taught me how to serve, and not expect anything in return, and yet how to be a steward of much.

It taught me about banking, and that the Bank of Heaven cannot be broken. It taught me there are three things necessary to draw on that bank: One, that I must be born into the Kingdom; two, that I must make deposits (my tithes and my works); three, that I can draw as much as I had deposited if I had enough faith.

The Bible can show you how all your needs and all your wants are supplied.

It can tell you what to do each day.

It can be the joy of your heart.

It can condemn you when you make a mistake until you repent. It gives you the desire not only to repent but to be perfect. "Be ye perfect," thus says the Lord.

It teaches you how to stand in tribulation.

It teaches you how to be a saint, an adult saint; and the saints are insult-proof.

And, greatest of all, it teaches you how to pray without ceasing, and how to love, how to be a lover. Because God is love, and the more you love, the more like God you are. This is the purpose of THE BOOK, to become like our Creator. To walk in the footsteps of His lovely Son, the Lily of the Valley, the Bright and Morning Star, the Rose of Sharon, the Lion of Judah who can break every chain and renew your faith, again and again.

I love the Bible because I love the Author, the One who inspired the writers of it.

The Bible taught me about angels. It taught me how to open my eyes that I can see angels. It taught me how to listen to the voice of the angels as they guide me.

It taught me how to see people who are humble, humble servants of God. And to say, "Thank you, Lord, for those servants."

It taught me how to pray for people.

It taught me patience.

It taught me how to forgive and forget.

It taught me that absolute forgiveness is absolute forgetfulness.

It showed me where I came from, why I am here, and where I am going . . . if I serve Him, if I faint not, if I make restitution for the mistakes that I make, if I ask forgiveness, and if I forgive.

It has made me fearless. I fear not what men can do to me, but I do have one fear. My greatest fear is that I will do something that the Holy Spirit will be removed from me!

Dear God, dear Father, take everything I have. Strip me and leave me nothing, absolutely nothing. Leave me friendless, leave me broken, leave me hungry, leave me starving, leave me ill, but remove not Thy Holy Spirit from me.

So, here I am, Lord.

Speak, thy servant heareth. You said that your sheep know Your voice. Continue to speak that I may know You better. I will listen; I will obey. I will be a good boy. Lord, how I need You! I can't live without You, Father. I need more of Your power. I need more of your Holy Spirit to strengthen me day by day.

Father, I thank you for the many friends that You have given me. Send me people that I can divide some of the love with that You have given to me when my cup overflows.

Thank you for letting me hear the heavenly music, the music of the spheres.

Thank you, Lord, for little kitty-cats that lick up the spilled milk. Thank you, Lord, for little puppy dogs that eat the crumbs that fall from Your table. Thank you, Lord, for letting me lick up the milk. Thank you, Lord, for letting me eat the crumbs that fall from Your table. Thank you, Lord, thank You, thank You! Dear Lord, may I leave some of the crumbs for others and not use them all myself. Amen.

Chapter 22

THE MAN CAREY A. REAMS

Working with Carey A. Reams on his book, CAREY A. REAMS, A "MOSES" FOR HEALTH, *I discovered a very unique and godly man. Therefore, as his friend and publisher I write the following:*

Mr. Carey A. Reams, the man nobody truly knows, however, some of his friends know him much better than others, and they never cease to be amazed at his counselling, knowledge, his recall of events and accuracy.

He is a man dedicated to God as fully as he knows how. He will not admit that he is humble. He says he does not know whether he is humble or not. He says he does not know whether he is a genius or not. He has no argument with those who wish to call him humble, or a genius. He defines genius as a person who knows more and more about less and less, until he knows everything about nothing, and he does not claim any such knowledge or wisdom.

Mr. C. Reams is a quiet man who hates publicity with a passion. No one can excite him in any way. At times he may be filled with righteous indignation and put on an act that he could chew you up, but in one second after it's over he is as calm as the Sea of Galilee after Jesus said, "Peace, be still." He is only filled with righteous indignation when someone tries to interfere with a command that God has given him to carry out.

He is a man very difficult to describe. He does not think as most people think. When asked why, he says that mathematics is a language and he thinks and dreams in mathematics, so before he can answer a question he has to express it in words so as to be understood. He often speaks in parables and likes giving examples. He frequently answers questions by asking questions. He is very quick to say when he does not know the answer.

He is a man of true love of his fellow man and he does everything he can to lighten someone else's load. "If I cannot lighten someone else's load, I will not make it heavier. If I cannot be a part of the answer, I will not be a part of the problem," he says.

Mr. Reams does not get excited when he is falsely accused. In fact, he smiles and says that he enjoys it. He tells that he loves trouble, and if he goes three days without any trouble then he starts some because nothing is accomplished as long as everything runs smoothly.

He calls the daily paper the devil's Bible. His name appears often in the devil's Bible, and when asked, "Do you know what the newspaper said about you?" he said, "Is my name on the front sheet?"

If you said, "No," then he said, "I don't want to hear it. If I don't make the front sheet, I don't want to be bothered about it. And I even care very little about what Satan says about me on the front sheet."

Reams also says, "When everything else in the world fails to make you ill, just read the daily newspaper, or get the morning news and if that doesn't make you sick, you are in pretty good shape. Or else, if that doesn't make you sicker, then there is hope for your recovery."

Reams is a superb teacher and gives God all the credit for everything that he does. He has given everything he has to God and says that no one can steal anything from him because he has nothing. If they take something that God has loaned him to use, they have taken it from God.

On one occasion someone questioned him about riding in a Cadillac with a chauffeur. He said, "That's not my Cadillac."

"Whose is it?" he was asked. "How come you are riding in somebody else's car?"

He said, "It belongs to the Lord. The Lord just lets me use it, and He also furnished me with a chauffeur."

He frankly says that his Father is so good to him He gives him all of his needs, and supplies all of his wants. He has gotten to the place now where he is afraid to ask God for anything because he believes that God will give it to him. He relates God is now often giving him things that he did not ask for, did not want, but yet the Lord said to him, "I want you to be the steward of this for me," to which Reams replied, "Yes, Lord. Show me how. Send me your people and give me the victory and I will give You the glory."

Mr. Reams fought for his country on foreign shores in World War II in the South Pacific. He was a Battalion Commander and Major and was made so on the battlefront, and before it could be confirmed through channels he was in a truck that was blown up. Completely blinded in one eye, he had his helmet shot off of his head from the inside and a bullet went around the side of his head, shaving off the hair just above his ear. Also a bullet went through his arm. The scar is still there today, it has never healed.

In regard to his instantaneous healing received at the Kathryn Kuhlman meeting he reports, "Jesus made the biological laws of physics, the laws of health, and He can change them any time He wants to for His glory and He can do it instantaneously, but God has only given me the gift of teaching people how to cleanse the temple."

He learned how to handle men and designate authority. He says he is not a business manager but an executive. He makes a distinction between the two. God, he says, "is not a businessman either, because there is no other God for him to do business with. Consequently, He's in the miracle business and He has made us soldiers of the battlefront to do battle for Him. If we are not doing battle for Him, for the Father and the Son, then we may be on the supply line supplying things necessary for those who are on the battlefront. God did not call everyone to become soldiers, but he called upon all men to take up their cross and follow Him." This Reams has done.

Carey Reams is not only a chemist, mathematician but is also a biophysicist. He deals with chemicals with the expert ease that is as skilled as a skilled mechanic uses tools. He can read equations and pronounce long chemical words as easily as a first-grade teacher can read first-grade books.

Reams loves children very, very much, as well as adults and

senior citizens. Children do not quickly take to him. They seem to be afraid of him because of the word "doctor," but very quickly he puts the child at ease and they become great friends.

He is also a farmer and agricultural engineer. He has owned an agricultural engineering firm, doing the impossible in agriculture all over the free world. Reams retired in 1968 and had intentions of going into research. He maintained the laboratories, the mineral and research division, and gave the business to the other engineers. His aim when he was a young man was to be a medical doctor, but the great Depression came and he was not able to get into medical college. He then decided to study foods and food chemistry to help prevent people from becoming ill. He was surprised that most farmers did not care whether or not their foods made people sick or kept them well. Their remark was, "I do not grow foods to eat. I grow them to sell."

Carey Reams is a very good musician. He does not sing, but he plays a number of instruments, including the bassoon, trombone, accordian, baritone, alto horn, french horn, piano and organ. He is not a classical player, but does play well enough to get through a service. He does not have time enough to stay in practice on any of these instruments, but is a lover of music.

He is a lover of the great outdoors, and a fisherman. Plus being a camera enthusiast. He loves to take pictures of landscapes and flowers, and has many albums of pictures taken in various parts of the world.

He has owned citrus groves and been highly successful as a farmer. He knows food processing and makes the statement that much of our prepared foods today (canned and frozen) are 40 years behind of what science has developed for the preparation of higher nutritional foods. In order to prepare a higher nutritional food it must begin before the seeds are planted, or before the trees come into blossom.

Reams is an excellent cook, and does not mind washing the dishes. If he is ever a visitor in your home, you will find that he makes his own bed, cleans the tub, keeps his room tidy, and you hardly know he's there. He realizes the heavy load the new mother has and does all in his power to lighten that load. He is more like a member of the family from the first day, and it seems you have known him forever. He does not seem a stranger, and he does not "put on the dog."

He is a very fine public speaker. On one occasion he was asked to speak in a university in which half of the students didn't want him to speak, and began to boo. As he stood to speak he waited a number of minutes for the booing to die down and it didn't. Very loudly he said, "Would you like for me to tell you about the time I was caught in bed with another man's wife?" A few in the front rows quieted down. After a second and third repeat of the same question most of the students were sitting on the edge of their seats. Then he said, "An honest confession is good for the soul."

He left them in suspense for a bit then said, "You know what that man said to me?" The audience could hardly wait to hear what the man said to him. "I was five years old and my father said, 'Son, go get in your own bed. Your father is tired from a hard day's work and wishes to go to bed.' " It is remarkable the wisdom that God has given this man.

Carey is a man of great patience but says the best of saints have very sharp horns because they don't use them very often. Horns mean power; power with God and power with man.

C. Reams is a pilot of small aircraft. He became a pilot in 1936 and has continued flying until recently when his eyesight is not sufficient to either drive a car or to fly a plane without someone with him. He says that when he is flying he is "chicken." He believes in obeying every rule of aerodynamics and that God meant what He said when He said, "Tempt not the Lord thy God" (especially when you are in the air flying light aircraft!).

Recently while teaching a class on the Reams Biological Theory of Ionization he received word that his mother had died. No one in the class had even suspicioned the message he had received, but by the next morning the word was out that his mother had passed away the previous afternoon. Before the class began sympathy was expressed to him. He did not attend the funeral but continued his Father's business and said, "Let the dead bury the dead." He knew his brothers and sisters could take care of it without his being there. He had taken care of his mother for the last few years of her life. After eulogy was expressed in the class he said, "If it were not for Jesus today I would be an orphan."

He never shed a tear. The class shed the tears for him.

A few days later he received the bill from the funeral home. With the check he sent a note: "Thank you for delivering my mother to the angels, to be taken to the throne of my Saviour."

He does not perform his religion, he wears it. He thinks it. He breathes it. He talks it. You will not be around him very long before he will be telling you something about the greatness of God.

Often when he lectures, Reams will say to the audience, "There is one book on health I highly recommend, please get out your pencil and paper and write the name of this book because I don't want you to forget it." He waits for them to get their pencil and paper then says, "This is one book that I have never found to be wrong, and it's absolutely accurate. It will work every time, if you will follow the instructions." By this time the audience can hardly wait for the name of the book. Then he says, "The name of the book is the Bible."

This man is impartial to race, religion, creed and sex. He treats all alike. He has drawn a circle and taken in all people of all nations and loves all, and says, "Anyone can love the loveable, but it takes an adult saint to love the unloveable, and when you can do that you know that you have been through the fire of tribulation." He does not condone sin, but sees a saint in every sinner. He sees a healthy person in every sick person. It is his desire that America will become the healthiest nation on earth, receive this health message and extend it to our children in the schools. It is his desire to see the hospitals emptied, except for those persons that harm themselves, or persons who are accidentally injured, or for childbirth.

He believes that women should be in a hospital to bring forth their children because there are many things that could happen and there a physician has everything to do with that is necessary.

He believes that surgeons do a very remarkable job. He finds most good surgeons and physicians do the best they have been trained to do, but know very little about tailor-making diets for individuals.

Reams believes in surgery, but he believes in using surgery as a means of last resort. During the war he was blown up in a truck and his right kidney was torn out and had to have surgery a number of times to get this kidney back into place. He also had to have 177 pieces of shrapnel taken out of his body. One piece went through his pancreas and to this day the pancreas does not perform its normal function. He has to take a certain medicine (it is also considered a food), which is and could be used as a drug, if used unwisely. As long as it is used with temperance, it permits his body to stay normal.

He is one of the six soldiers of World War II whose temperature went below 90 degrees and he still lived. His went down to 76 degrees for a number of months and he was almost freezing to death when everyone else was just dying of heat and sweat in the South Pacific. This pancreas is something he has to live with, and he handles his problem well. You could be around him for months and not know that there was anything of this nature still giving him a problem. He does not discuss this unless someone happens to be around when he needs to take the medicine.

He says, "I have been arrested more times than the Apostle Paul. I have not stayed in jail as long as Paul but I have been falsely accused every time. I have never broken the laws of our country," then adds, "I thoroughly enjoy being falsely accused by Satan, for we are told in Revelation 12:10 that Satan is the accuser of the brethren." It deeply amuses him when the hounds of hell blacken him because, "Then I know I'm on the King's highway. When I go three days without hearing the hounds of hell barking I become afraid that I'm on a detour."

"I never cease to be amazed," says Reams, "how eager some people are to get to the cemetery. They just can't wait. They break every rule in the book. They smoke, drink harmful liquor, eat the wrong way, never take a day's rest, always so busy waiting on themselves until they simply burn themselves out. This type of people never grow old, they always die young. They are never happy. They are miserable." He also comments how tightly some people hold everything they have—with a clenched fist. "You cannot receive the blessings of God unless you open your hand and receive. You can only keep that which you give away. You cannot take money to heaven with you. The one question that is going to be asked you when you stand before the great white throne of God (if you stand there, many will not stand there, and those who do it will not be to determine whether or not they are going to be in heaven or hell. It is going to be to determine your rewards): 'Whom did you bring with you? Where is that pretty flock that I gave you?' " (Jer. 13:20).

Reams appreciates good, clean humor, as well as jokes on himself.

He laughs with the Amish people who say that his clinics or retreats are places where people get well on lemonade and stale jokes. They have named the streets of the retreats such as Lemonade Lane, Prune Juice Circle, and Green Drink Lane, etc.

One bright May day we went into a restaurant in Tampa, Florida. He had on his driving glasses. In the restaurant they had beautiful white tablecloths and linen napkins, plus handsome waiters. They were dressed in long-sleeved, cuffed white shirts, black vests, white pants with a black stripe down each side. The waiter brought a glass of water, then a many-page numbered menu. There was a pencil and little pad hanging on the menu. All one needed to do was to write the number of the menu on the little pad. The pad also had the table number and chair position. Suddenly Reams realized he had his driving glasses on instead of his reading glasses. He said, with a big smile, to the waiter, "Would you mind reading this menu for me?"

With his big white teeth showing, this dark-skinned waiter said to him, "I'm awfully sorry, boss man, but I'm just as ignorant as you is!"

The honesty of this young waiter tickled him. He left the restaurant, got his reading glasses and then read the menu.

He thoroughly enjoys other people disagreeing with him because "where you and I disagree, there one or the other of us will begin to learn," he says.

He finds many people are afraid of progress and have enmity toward the Biological Theory of Ionization. "If more people loved each other this world would be a better place to live in. This nation is dying for leadership, men who are statesmen that can become heros of our youth, and dying for want of love for their fellow man.

"Churches are something like grades in school. When you get all that church has to offer, then you should become so religious until you get kicked out. This is the way God promotes you. You will not ever amount to anything until you get kicked out of at least six churches. Then you will know that you are growing up. You will know then that your life is beginning to become pleasing to God."

One thing that simply irritates Reams is the harm that is being done in the Kingdom of God by beggar preachers that drive old cars, live out of the missionary barrel and are all the time begging for their daily bread from man and then think that they are holy. "What rich sinner wants to be like this poor preacher? Or, what healthy sinner wants to be like a sick saint? God wants His people to be examples of His Kingdom and to exalt Him in word, possessions, deeds and actions. There are many church people who think wealth is a disgrace. Look at Moses. He was one of the highest educated men in the whole earth, trained to become the prime minister of Egypt. Then God took him out to the wilderness for 40 years and taught him how to be and how to do things the way that God wanted them done. What God commands us to do often does not make any sense. However, we should do it whether or not it makes any rhyme or reason. It does not make sense to lead 300,000 Jews from the land of Egypt, across the Red Sea, through the wilderness and into the land of Canaan; yet, God did it. And when they got to the land of Canaan (which is only about a six-hour automobile drive today) they were afraid to go in. They refused to obey Moses and to go into the land. So God let them march in the wilderness for 40 years and have nearly 298,000 old-fashioned funerals by letting all the old persons die. Then God led them to the land of Canaan. By this time Moses became a bit over-enthusiastic because this time the young were ready to go into the land. God said to Moses, 'Speak to the rock and let water come forth.' Instead of speaking he struck the rock. Then the Scripture tells us God buried him in the land of Moab and no man knows the whereabouts of that grave.

"Moses learned in 40 years absolute obedience and this is one of the greatest tasks for us today. In order to learn absolute obedience, we must learn absolute forgiveness, and absolute forgiveness is absolute forgetfulness. Faith is absolute expectation. And faith without works is dead. The weaker and sicker we are, the less our works. The healthier we are the greater our works."

This is the philosophy of Carey Reams. He believes if more of the pastors would feed the sheep instead of teaching them history, churches would be filled and running over with noisy, hungry sheep.

Mr. John C— one day was introducing Carey Reams to an audience and said the Reams Biological Theory of Ionization for perfect health is the greatest message to mankind since the word was proclaimed to the world that Jesus had arisen.

"It is the most neglected message in the entire Bible," says Reams. He believes that it is the message that will unite the saints for the coming of Jesus. He believes and lives what he preaches and does it without fear.

He has a rather peculiar philosophy. He said it's a day ruined when somebody doesn't give him a piece of his mind or chews him out. However, he is afraid to give anyone a piece of his mind because he says he doesn't have that much to spare.

He tells the story that he and his wife had not been married very long before he did some foolish thing and his wife was really letting him know that he was not the king that she thought she married. He said, "I'm sorry, dear. Forgive me. If it were not for you I might not make it through the pearly gates." And that woman could not stay mad at a man who said that.

Another time he did something silly and deserved a good reprimand. His wife said to him, "What would you do if I left you?"

"Honey, you know what I would do if you left me?"

"No," she said, "what would you do?"

"I would go with you."

She flat could not stay mad at him. They are the two most charming love birds you've ever seen.

His wife adores him, and so do his children, his secretary, and employees. No man knows a man any better than his own children, his secretary or his administrative assistants, and yet there's nothing they would not do for him.

Some have asked Carey Reams about his scholastic background. This is his answer:

"Is a man any better mechanic because he bought his tools from any special warehouse?

"I studied in some of the world's best universities and under famous tutors. My wisdom came from God. He is my best teacher.

"At my age I do not care to tell where I got my tools lest I should boast of any other than God alone. Education is only the tools anyone works with. Wisdom comes from God. Without knowledge and ordinary horse sense, or common sense, anyone would not know what God was saying to him in higher mathematics, chemistry, physics, biophysics, or any other related field unless he had prepared himself educationwise by studying to show himself a workman that needeth not to be ashamed, rightly dividing the truth. God said, 'To him who has it shall be given.' I praise God that He has chosen a nobody like me to help carry the cross."

Reams is a man of great faith. He does not tell God how to answer his prayers. He seldom ever asks for anything. He just knows that God will supply his needs and his wants for His glory. He never screams at all. He never says anything against those who scream even though he often tells them that there is a better way of doing things. Screaming isn't faith. Children of God may have to scream until they learn to exercise their faith. But once they learn to exercise their faith, then all things are theirs, he says.

"One of the sad things I have witnessed is the people that are deathly ill go to a religious meeting and come out saying, 'Praise God, I am healed,' and our instruments show they are not healed. And yet they will not believe these instruments; consequently, they die. It is glorious for people to go to a meeting and be healed, and when they are healed, these instruments that the Reams testers use also show they are healed," comments C. Reams.

He is the first man to ever express every degree of body chemistry in biological mathematical terms. Reams hates the word "diagnosis" because he wants no one guessing at anyone's life. He says he has never made a diagnosis and he never will. He says he does make an analysis which can be proven and substantiated. He invites a true investigation of his methods and is willing to teach it to those that are able to use it. He wants to use it only for the glory of God.

Reams is often accused of a conspiracy against the nation. He does teach seminars, and each tester goes out and does his thing in his own way. The same as those who take a Dale

Carnegie course. He does not any more have a conspiracy than any university has which teaches in the chemistry classes how to manufacture or distill whiskey, or how to make wines. This does not make the students a conspiracy against the nation. There is no way that he is linked directly to the students, nor does he receive anything from the students other than the price of the seminars.

There have been three attempts to poison Carey Reams. One was on October 3, 1975, by putting strychnine into a pear he had on his desk. He took one bite of it and in fifteen minutes was desperately ill. He heaved for several hours. He knew not to drink any water. He knew not to go to a hospital in that area because the doctors would have given him something to kill him.

On January 23, 1977, poison was placed in his food while he was in one of the larger motels. He was in the hospital 89 days, and as a result of this poisoning had to go through surgery twice. Once while in intensive care the oxygen was turned off and the administrative assistant came in, found him in a coma. It was 14 hours before he came back to reality.

He has to be careful where he travels, where he eats. Most of the time he travels by car with a bodyguard. It is strange that in a free country a man like Carey Reams has to take all these precautions to live. There are countries in the world which offered him all kinds of protection if he would come there and practice, and would even give special air rate for people coming to see him, and yet he has not accepted these offers because he believes that America is worth fighting for.

When asked how many grandchildren he has he says, "Fifteen, the last I counted them." He wants America to be a place where his grandchildren can have life, liberty and freedom of speech and freedom to go to whom they wish with their health problems, without needing attention by the police, or without harassment by the government. He wants them to have a freedom of choice to choose their work. He wants them to grow up to be men and women of God. Persons who are fearless, and will carry on regardless of the problems that are put before them.

"I have no troubles and no worries," remarked Reams. "Trouble is an assignment, a problem, that God gives us to cuddle us a little closer to our heavenly Father. Problems become trouble when a person with the assigned problem is either in too big a hurry to work at the problem, or feels that they are too good to work at the problem, or are so panic-stricken that they refuse to tackle the problem, or else they are just too lazy to do it, or they can procrastinate, then the problem becomes a worry.

"So these are the conditions that make for worry. I love my wife very much but for 25 years of our married life she stayed two weeks behind with her worrying. I felt so sorry for her. One day I asked God about worry, and He said, 'WORRY IS THE DEVIL'S PRAYER.' So I wrote on little cards: JESUS SAID THAT WORRY IS THE DEVIL'S PRAYER, and stuck them up all over the house. And to this day she has never worried again. It is absolutely impossible to worry and to pray to the Lord Jesus Christ at the same time."

Reams is a very unselfish man and anything that he can do for anyone he does and he does not feel that he is being imposed upon. Most of the time he will do things before he is asked. He is very appreciative of any kindness extended to him.

One of the things that he said was a little difficult for him was that when younger people wish to help him across the street, or to do anything for him, he had to learn to accept it graciously, even though he felt he did not need it. He was a Boy Scout (Eagle Scout and Scout Master) many years and loves to work with young people. He is never too busy to help them with their problems.

Reams has some favorite sayings such as:

"Hate is one of the great causes of illness, beside the mineral deficiency."

"Hate is the breaking of the Golden Rule because you must hate yourself before you can break the Golden Rule."

Regarding the Golden Rule Reams says, "Much illness could be prevented, food digestion improved and constipation greatly relieved if more people would live the Golden Rule. It is impossible to speak unkindly about anyone and love them too. Anyone may unintentionally cause a person to feel inferior or may be a threat to his vanity, and consequently, in his own egotistical defense he forgets the Golden Rule.

"In all cases that this situation exists the predominate factor is fear. So in self-defense unkind words are spoken.

"Anyone who disobeys the Golden Rule only exposes his own true character. ('Judge not, that ye be not judged. For with what judgment ye judge, ye shall be judged: and with what measure ye mete, it shall be measured to you again. And why beholdest thou the mote that is in thy brother's eye, but considerest not the beam that is in thine own eye?' Matthew 7:1-3. 'But we are sure that the judgment of God is according to truth against them which commit such things,' Romans 2:2).

"No one can break the Golden Rule and be truly humble. Humility is hidden from the eyes of its possessor.

"No one has a right to judge another. God has reserved this position for Himself. Only God has the power to destroy the soul, even after death.

"We do have a right to be a 'fruit' inspector, providing we do not permit anyone to know that we may be challenged to be his fruit inspector. The second we crow about the bad results of our fruit inspecting we become a critic and a judge. The judgment of character by another is a sin. It matters little whether the judgment is just or unjust. The Golden Rule was broken the moment the first unkind word was spoken. Such action affects our health.

"Never push down anyone. If you push him down you must go down with him. Lift him up and you will go up with him.

"There are many ways to lift others up and evaluate their fruit without transgressing the Golden Rule. You can say things that will not degrade and not break the Golden Rule and hopefully lift anyone up. At least no evil is thought and no harm done. You may say, 'He does very, very well considering his mentality. Except by the grace of God I could be just like him.' This shows love. Or, you may say, 'That fellow is a genius. His most serious handicap is his handicap.' Again love is shown. Again, you can say, 'That fellow does the best he can. He probably never had the opportunities that you and I had.' Or, you can say, 'Just because he is different please do not berate him. Some very great men were considered to be morons. God loves all His creatures, even morons. May the Lord bless him.' You might also say, 'I admire his spunk even though his dreams seldom come true. It is better to die trying than never try at all.'

"So if you want better health, lift up your fellow man."

Another of Reams' favorite sayings is: "There's no money in gossip. One of the reasons that so many people are so broke all the time is that they are so busy attending to other people's business until they don't have time enough to work at the job that pays well for their time and their effort."

He also says that "Dreams are only profitable when made into a blueprint and the blueprint into building the ideal of your dream, and this is better done when fertilized with faith, cultivated with patience and with sweat and elbow grease to bring the blueprints into reality."

Two of his favorite poems are:

>If you can keep your head when all about you
>>Are losing theirs and blaming it on you,
>If you can trust yourself when all men doubt you,
>>But make allowance for their doubting too;
>If you can wait and not be tied by waiting,
>>Or being lied about, don't deal in lies,
>Or being hated, don't give way to hating,
>>And yet don't look too good, nor talk too wise:
>
>If you can dream—and not make dreams your master:
>>If you can think—and not make thoughts your aim;
>If you can meet with Triumph and Disaster
>>And treat those two imposters just the same.
>
>If you can make one heap of all your winnings
>>And risk it on one turn of pitch-and-toss,
>And lose, and start again at your beginnings
>>And never breathe a word about your loss.
>
>If you can talk with crowds and keep your virtue,
>>Or walk with Kings—nor lose the common touch,
>If neither foes nor loving friends can hurt you,
>>If all men count with you, but none too much;
>If you can fill the unforgiving minute
>>With sixty seconds' worth of distance run,
>Yours is the Earth and everything that's in it,
>>And—which is more—you'll be a Man, my son!
>>>—Rudyard Kipling

Out of the night that covers me,
Black as the Pit from pole to pole,
I thank whatever gods may be
For my unconquerable soul.

In the fell clutch of circumstance,
I have not winced nor cried aloud:
Under the bludgeonings of chance
My head is bloody, but unbowed.

It matters not how strait the gate,
How charged with punishments the scroll,
I am the master of my fate:
I am the captain of my soul.
—William Ernest Henley

Carey A. Reams loves the Bible most of all. He cherishes it, feasts upon it and lives it to the best of his knowledge and ability.

This man will help you if you are sick regardless of all the threats that are made against him. He is a man who is at peace with his God and one of his favorite expressions is: "Perfect love casteth out all fear." There are three things necessary to share God's love and His glory, says Reams, namely:

"1. Give yourself and put all your possessions at His feet.

"2. Completely empty yourself of yourself, cling not to worldly goods, turn loose of all your pride and self esteem, become as a little child that is absolutely obedient to his parents, be thankful for His tender mercies and praise Him for all problems He assigns you to strengthen your faith. Rejoice in all tribulation for His name's sake.

"3. Keep His commandments and teach others so that your reward will be great in His kingdom. Then ask whatsoever you will in His name and He will answer your prayer, if you will take your prayer request the way He sends it."

Reams' message to drug pushers and drug salesmen is: "USE DRUGS AS LITTLE AS POSSIBLE. USE THE WORD OF GOD AS MUCH AS POSSIBLE. GOD IS READY TO DELIVER YOU FROM POOR HEALTH INTO GOOD HEALTH WITHOUT DRUGS, OR WITH A LOT LESS DRUGS."

Since he made these statements his nickname among the students of the seminars that he teaches is "Moses."